Organizational Issues in High Technology Management

MONOGRAPHS IN ORGANIZATIONAL BEHAVIOR AND INDUSTRIAL RELATIONS, VOLUME 11

Editor: Samuel B. Bacharach, Department of Organizational Behavior, New York State School of Industrial and Labor Relations, Cornell University

MONOGRAPHS IN ORGANIZATIONAL BEHAVIOR AND INDUSTRIAL RELATIONS

Edited by
Samuel B. Bacharach
Department of Organizational Behavior
New York State School of Industrial and Labor Relations
Cornell University

Volume 1. **ORGANIZATIONAL SYMBOLISM**
Edited by Louis R. Pondy, University of Illinois
Peter J. Frost, University of British Columbia
Gareth Morgan, York University
Thomas C. Dandridge, SUNY, Albany

Volume 2. **THE ORGANIZATION AND ITS ECOSYSTEM: A Theory of Structuring Organizations**
Charles E. Bidwell, University of Chicago
John D. Kasarda, University of North Carolina

Volume 3. **MILITARY LEADERSHIP: An Organizational Behavior Perspective**
David D. Van Fleet, Texas A & M University
Gary A. Yukl, SUNY, Albany

Volume 4. **NATIONALIZING SOCIAL SECURITY IN EUROPE AND AMERICA**
Edited by Douglas E. Ashford, University of Pittsburgh
E.W. Kelley, Cornell University

Volume 5. **ABSENTEEISM AND TURNOVER OF HOSPITAL EMPLOYEES**
James L. Price, University of Iowa
Charles W. Mueller, University of Iowa

Volume 6. **INDUSTRIAL ORGANIZATION: A Treatise**
Joe S. Bain, Professor Emeritus of Economics, University of California, Berkeley
P. David Qualls, University of Tennessee, Knoxville

Volume 7. **STRUCTURING IN ORGANIZATIONS: Ecosystem Theory Evaluated**
Charles E. Bidwell, University of Chicago
John D. Kasarda, University of North Carolina

Volume 8. **PUBLISH AND PERISH: The Organizational Ecology of Newpaper Industries**
Glenn R. Carroll, University of California, Berkeley

Volume 9. **ORGANIZATIONAL CHANGE IN JAPANEESE FACTORIES**
Robert M. Marsh, Brown University
Hiroshi Mannari, Kwansei Gakuin University

Volume 10. **STRATEGIC MANAGEMENT FRONTIERS**
Edited by John H. Grant, University of Pittsburgh

Volume 11. **ORGANIZATIONAL ISSUES IN HIGH TECHNOLOGY MANAGEMENT**
Edited by Luis Gomez-Mejia, Arizona State University
Michael W. Lawless, University of Colorado, Boulder

Volume 12. **STRATEGIC MANAGEMENT IN HIGH TECHNOLOGY FIRMS**
Edited by Michael W. Lawless, University of Colorado, Boulder
Luis Gomez-Mejia, Arizona State University

Organizational Issues In High Technology Management

Edited by: LUIS R. GOMEZ-MEJIA
Arizona State University

MICHAEL W. LAWLESS
University of Colorado

Greenwich, Connecticut *London, England*

Library of Congress Cataloging-in-Publication Data

Organizational issues in high technology management / edited by Luis R. Gomez-Mejia, Michael W. Lawless.
p. cm. -- (Monographs in organizational behavior and industrial relations ; v. 11)
Includes bibliographical references and index.
ISBN 1-55938-104-3
1. High technology industries--Management. 2. Corporate culture. 3. Organizational behavior. I. Gomez-Mejia, Luis R. II. Lawless, Michael W. III. Series.
IN PROCESS (ONLINE) 90-38799

55 Old Post Road No. 2
Greenwich, CT 06830

JAI PRESS, LTD.
118 Pentonville Road
London N1 9JN
England

ISBN: 1-55938-104-3

Library of Congress Catalogue Number: 90-38799

Manufactured in the United States of America

CONTENTS

INTRODUCTION

Much attention has been paid in recent years to the so called "high tech" revolution in both the mass media and academic publications. The unprecedented growth in the high technology industry during the previous decade is viewed by many as a crucial contributor to the economy by generating a large number of "desirable" jobs, introducing a plethora of new consumer goods, increasing labor's productivity and strengthening the U.S. competitive edge in world markets. While few would agree on a precise definition of a high technology firm, there seems to be general consensus that scientific discovery, innovation, and research and development activities are of paramount importance to these organizations. In fact, one often hears the label "knowledge intensive industry" as synonymous with "high tech."

Practitioners and scholars alike seem to agree that high technology firms call for a management style different from traditional companies in order to be successful. These firms must deal with a unique set of problems that require special solutions. Some of the management challenges facing high tech companies include: attracting, retaining and motivating R&D talent in the midst of severe industry-wide turnover rates; fostering employee commitment in work situations that are uncertain, unpredictable and highly unstable; creating a culture that supports flexibility and innovation alongside management control systems; designing organizational structures that provide some degree of order yet can be easily molded in response to rapidly changing conditions; managing the integration and differentiation of highly specialized subfunctional areas; and meeting the needs of R&D employees who value autonomy, intellectual pride and the pursuit of scientific endeavors that may not always be in the best interest of the firm.

This book brings together in one volume the thoughts of leading management scholars who have studied those organizational issues. Sixteen chapters, divided into four major sections, examine different aspects of the internal management process of high technology organizations. A brief overview is provided below. A separate volume to be published by JAI Press Inc. (edited by Michael Lawless and Luis R. Gomez-Mejia) will focus on the strategic and environmental forces impacting high technology firms.

Part I is concerned with the organizational culture of high technology firms. Five chapters deal with those issues. The first chapter, by Jelinek and Schoonhoven, report the findings of a longitudinal study of five large high technology firms. According to these authors, control of innovation in high tech firms takes place primarily through organizational cutlure rather than bureaucratic processes. Jelinek and Schoonhoven found that managers operating in these firms face an ongoing tension between stability and change, past commitment and future opportunity, discretion versus direction, delineation of responsibility versus fluid tasks, and close monitoring versus flexible supervisory systems. The ability to manage these conflicting forces is a key element in the eventual success or failure of high technology firms. The second paper in this section, by Beyer, emphasizes the crucial role that continuous change plays within the high technology firm. Beyer poses the question: "Can you develop commitment in the workforce and a culture of commitment in situations characterized by change, unpredictability of events and fluidity of work roles, group or industry?" She argues that fostering employee commitment demands that the firm adapts to cultural diversity of different occupational groups. According to Beyer, the more the firm can accommodate itself to diverse internal constituencies, the more committed employees are likely to be.

The next chapter in Part I, by Fryxell, focuses on the values, norms and belief systems of innovative groups. He labels these cultural dimensions "shared cognitions." After reviewing several streams of research dealing with these issues, Fryxell concludes that innovative high technology firms are faced with a "dialectic" value system, making it difficult for managers to handle the many conflicting values. Some of the contradictory values cited by Fryxell include: creative freedom vs. control, differentiation vs. integration, individualism vs. collectivism, localism vs. professionalism, and contemplation vs. action. The fourth paper in Part I, by Baba, argues that most innovative activities in high technology firms take place among informal work groups and networks. The end result of this informal process is what she terms "local knowledge." Baba develops a set of propositions around the central theme that "advanced technology organizations should encompass informal bodies of knowledge that are proportionately more complex and rapidly evolving than those located in traditional organizations." The last chapter in this section, by Zahrly, argues that myths, values and ideology play an important role in the successful management of high technology firms. Because employees tend to create their own idiosyncratic "social enactment of reality," it is incumbent upon high tech management to provide a common meaning within the organization. Zahrly believes that most successful high tech managers are able to develop a shared ideology in the workforce in support of creativity, teamwork and free flow of communication. In her chapter, Zahrly describes the results of a study in a start-up high tech firm that seems to support this notion.

Part II deals with political processes and the management of growth in high technology organizations. The first paper in this section, by Page and Dyer, examines the political process involved in radical, discontinuous product/process innovations leading to dramatic technological breakthrough. According to these authors, radical innovations may threaten the status quo and are likely to be resisted by entrenched groups inside and outside the firm who may have a vested interest in the current technology. Therefore, according to Page and Dyer, successful radical innovation demands effective handling of political forces and not just the technical dimension. This requires obtaining sponsorship for the project from top management, building coalitions in the organization and gaining the support of threatened groups. Sexton's paper, the second in this section, is concerned with the management of growth in high technology firms. He argues that growth in high technology firms does not follow a predetermined metamorphosic curve as implied in widely held life cycle models of organizational growth. According to Sexton, firm growth is not a natural phenomena, but rather socially determined, resulting from a conscious decision made by a given CEO/entrepreneur. The observed rate of growth in high technology firms depends on the entrepreneur/CEO's "propensity for growth," which is in turn a function of his/her psychological profile. The extent to which growth is successful, however, depends on the CEO/entrepreneur's ability to manage it.

Part III of the book is concerned with organizational interventions designed to enhance the performance of high technology firms. The first chapter in this section, by Carroll, describes a "multiple management approach system" (MM) designed to improve existing products and services, develop basic management skills and melt barriers between different organizational subunits. The MM system involves the convening of cross-functional middle management teams who meet on a frequent basis to discuss issues of common interest to the company. Carroll illustrates how this organizational intervention has worked in several firms. The second chapter in this section, by Warrick, argues that the rapid rate of change in high technology firms places a premium on two skills: ability to successfully manage change, and ability to develop a high performance organization under conditions of extreme uncertainty. He provides a step-by-step approach for organizational development aimed at improving those two skills. His five target areas include: leadership, organization structure, teamwork, workforce utilization and employee involvement. The last chapter in this section, by Kolodny, describes several organizational forms available to organizational designers that are congruent with the nature of high technology firms. According to him, "technological environments will almost always dictate organizational forms, but more as 'envelopes' of structural forms within which a variety of processes and approaches are possible."

The last section, Part IV, is concerned with the management of human resources in high technology firms. The first chapter in this section, by Cascio, provides an overall perspective of how human resource management (HRM) may be used strategically in high technology companies. Cascio reviews the crucial contingencies affecting these firms and policy choices available in each of the major HRM sub-functions, including staffing, compensation, training and development and labor relations. Cascio concludes his chapter with a case study of HRM practices at Hewlett Packard in Singapore. In the second paper of this section, Hybels and Barley argue that HRM practices that are presumably unique to the high tech industry (e.g., flexible time, open door policies, stock ownership) originated in other industries and typically only apply to a small professional elite in high tech firms (mostly R&D workers). What sets these firms apart, according to them, is the extent to which those practices are widespread and uniform in the industry. Hybels and Barley attribute this shared HRM practice in the high tech industry to supply imbalances in the labor market, particularly for R&D workers. Given a tight labor market, much imitation occurs among high tech firms in order to provide equally attractive settings to potential recruits. The third chapter in this section, by Bretz and Dreher, is concerned with the match between individual and organizational characteristics in high technology firms. They argue that the type of person who would be attracted to the high technology industry is one who is achievement oriented, willing to incur risk and has a strong desire for autonomy. This individual, according to Bretz and Dreher, would be happier in a work environment characterized by strong emphasis on ability and achievement, competition for promotion and raises and the use of past accomplishments as criteria for personnel selection. Such an employee will not respond well to a management philosophy whose goal is to create a high degree of commitment and loyalty to the firm.

The fourth chapter in Part IV, by Turbin and Rosse, is concerned with the staffing problems faced by high technology firms as a result of the high attrition rate of S&Es. They review the factors responsible for these problems and the consequences to the firm. Their chapter concludes with an agenda for future research in this area. The fifth chapter in this section, by Ferris and Buckley, is concerned with evaluating the performance of R&D professionals in high tech firms. They argue that we know little about how appraisals are conducted in this industry, and factors moderating their success. Ferris and Buckley present an exploratory study to address those issues, based on a sample of supervisors in 32 high tech firms. They found that objectives-based systems (e.g., management by objectives) were most often used, followed by trait-based rating scales. Suprisingly, peer review was nonexistent. They also report widespread dissatisfaction with the appraisal system in these firms. The sixth and final chapter in Part IV, by Gattiker, develops an interdisciplinary model that links together technological innovation, organizational adaptation, culture and employee training. The central thesis of his model is that successful technology adaptation, organizational adaptation and employee training require a match between organizational culture, national culture and the cognitive map of employees.

Luis R. Gomez-Mejia
Michael W. Lawless
Editors

PART I

ORGANIZATIONAL CULTURE OF HIGH TECHNOLOGY FIRMS

MANAGING INNOVATION IN HIGH TECHNOLOGY FIRMS:
CHALLENGES TO ORGANIZATION THEORY

Mariann Jelinek and Claudia Bird Schoonhoven

LESSONS FOR MANAGEMENT FROM HIGH TECHS

Innovation management is complex, although typically its aim is simply stated: to keep innovation happening effectively. The complexity comes from a variety of constant contradictions and tensions that must be held in balance through culture, norms, attitudes and understandings, and above all, through managers' attention. These tensions reflect genuinely conflicting demands in the area at the heart of successful innovation strategy; they are the source of its vitality. They come together most visibly in high technology firms.

High technology firms are not the only innovative companies, nor even the first companies to make use of the approaches we will outline. "Team management," "strong culture," even structural change, and many other aspects of the "high technology" innovation management methods are to be found elsewhere. If these firms are not unique, however, they do exemplify an especially challenging environment, where the tensions of managing innovation are especially visible.

The high technology environment is also one where many of the theoretical assumptions and findings of past organization studies seem to lose their potency, and to mislead more than they assist us in understanding these firms and how they operate. Indeed, we shall argue that much older organization and behavior theory is inadequate in the context of persistent innovation amidst control that characterizes these companies. A number of contemporary researchers have begun to address the problems of innovation management (e.g., Burgelman & Sayles, 1986; Daft, 1986; Kanter, 1983). Unfortunately, others still often cite uncritically, data and conclusions from much earlier, simpler times, with all their assumptions intact.

This paper reports the conclusions of a study of high technology firms carried out over the past eight years involving five sizeable firms and over 100 managers who were interviewed at length. Necessarily, the complete logic of our argument is not reproduced here. However, we believe its outlines are clear. In the following discussion, we will highlight the tensions amidst which high technology firms operate, and the theoretical and practical questions we see arising from their experience. We will begin with some key innovation management questions.

STABILITY AND CHANGE

Semiconductor electronics firms offer a particularly cogent example of innovation challenge, and of a peculiar, enduring tension between stability and change. Stability is needed in high technology firms to manage the complicated production activities used to manufacture electronic devices. Each step of the elaborate production process is crucial: failure anywhere along the way—through perhaps hundreds of steps—typically results in failure of the entire device; you cannot rework integrated circuits. Stabilizing manufacturing processes in order to control them is a prime necessity.

Similar complexity characterizes the innovation process. While an initial design concept or new insight may be the work of a single person or a small team, its specific application in a manufacturable circuit design may well involve dozens or even hundreds of person-years of effort. Routinely today, specific circuit designs *require* computer assistance: they are too complex for unaided human accomplishment. Specifying the directions for the deposition of repeated layers of circuitry, the necessary connections in submicron sizes and the meticulous control of temperatures, gas concentrations and the assembly and packaging as well—all require a great deal of careful coordination to insure that each step in a lengthy, exceedingly finicky procedure will be precisely controlled. Ultimately, of course, design demands and manufacturing processes must mesh seamlessly.

As with many manufacturing processes, introducing change into semiconductor manufacturing means that any unexpected difficulties must be mastered to again reduce the process to control. Many constituencies have a stake in stability, and their legitimate needs can induce quite rational resistance to change.

- *Manufacturing's desire for stability in processes.* New processes, and, to some degree, new products, require mastery of changed activities; the greater the change, the more difficult it may be to return the process to control in order to achieve required yields.
- *Engineering/design's desire to serve external or internal customers.* He or she who understands the old approach has efficiency in using a prior design, or established design concepts, and will lose some portion of that efficiency with change.
- *Marketing's desire to understand technical specifics, to be able to describe them, understand how they assist customers and portray their salable capabilities.* Where product characteristics change substantially, new customers must be assimilated. Entirely new approaches to meeting customer needs may have to be comprehended and articulated. New customers may not behave as older customers do, and they may not be reachable through established channels.

Despite these resistances, innovation can happen, must happen and does happen repeatedly in high technology firms. Indeed, with some frequency, change goes well beyond even planned change, to embrace the quintessential opportunism of discovery and invention, not merely incremental improvements. Such opportunistic innovation underlines the difficulties of managing innovation in these firms. Plans and strategic commitments are essential, yet so too, is change. How can legitimate concerns for stability be balanced against the constant need for change?

Plans and Happenstance

Plans help insure coordination, yet creativity often dawns from making use of the opportune. High technology firms regularly do both: planning carefully, pushing hard to attain plans and sensitizing people to the benefits of chance opportunities that may upset plans. Despite the need for stability and the understandable resistances that we can predict, these firms exhibit extraordinary levels of continued innovation. Many product life cycles are quite frequently less than 18 months, and the "innovator's premium" of unchallenged market position devoid of competing alternative products very short indeed. Global competition in a technologically sophisticated marketplace, inhabited by numerous competent adversaries, necessitates repeated innovation. Nevertheless, it's quite clear that bringing new technologies into production, building a wafer fabrication line, or developing a new market can take years of persistent, sustained effort. How, then, can strategies be maintained over time?

Figure and Ground, Overview and Details

Effective management demands mastery of details in many businesses, not just in semiconductor electronics. In industry after industry, details make the difference between success and failure. It's not enough to have a general picture; the concrete specifics of the business—the ones that really matter—must be deeply understood at all levels of the organization, especially the highest. Even though senior managers in technology-based firms are typically "a long time away from the bench," they must ultimately make strategic commitments that determine the organization's fate. How can these firms insure that senior managers' decisions are technically adequate?

Yet senior management must not become wrapped up in the details or in focusing on minutia: senior management's special obligations are understanding the big picture and articulating that understanding. The role of senior management, their links with juniors and their dependency upon juniors for technical specifics are vastly different from roles and relationships as depicted in much research literature and management. There simply is not room here for the authoritarian "great leader," whose responsibility is "deciding strategy," or formulating it, or laying out the plan that others will implement. What, then, *is* senior management's role?

Herein lays a paradox: senior managers are responsible for strategic commitments, but cannot possibly acquire or maintain sufficient amounts of current, technically specific information to make those decisions by themselves. Although high technology senior managers are typically thoroughly trained in a technical specialty, and may have contributed to their firms as technical specialists during their careers, their role as senior managers requires a different perspective. They must transcend their limitations by opening the strategic process to others simply because the necessary wherewithal of technically specific insight, ideas and knowledge by far surpasses that possessed by even the greatest of leaders. It is clear that such leadership is not charisma, nor does it involve "participation," as that word is typically used. How can such leadership be usefully described?

CHALLENGES FOR THE FUTURE: INNOVATION AND CHANGE MANAGEMENT AS SURVIVAL STRATEGIES

Innovation *CAN* be maintained in large firms, as we see from the example of the high technology firms—but not by traditional bureaucratic methods. Those methods are too rigid, too unresponsive and too limited as mechanisms of motivation, coordination or control. Classical descriptions of the failure of bureaucracy (e.g., March & Simon, 1958; Morrison, 1966) are well known, and need not be rehearsed here. Yet in the firms we studied—Texas Instruments, Hewlett-Packard, Intel, National Semiconductor and Motorola—we did not find loose, "organic" management styles (Burns & Stalker, 1961), either. The degree of specificity, delineated responsibility, follow-up and order was surprising, the more so because innovation was so visible and pervasive a concomitant.

Neither did these firms meander strategically, despite their free flow of ideas and options. A "strategy du jour" is *not* appropriate where multi-million-dollar commitments to technology and capital equipment are the norm—but firms must also be prepared to change current strategy (as embodied in products, markets, technology positioning, the configuration of the firm and its mission). A review of the top ten merchant semiconductor firms' patterns of behavior during 1975-1979 (Schoonhoven, 1984, 1987) highlights consistency of strategic behavior; even a cursory review of the firms' current practices underlines significant changes.

For instance, Intel has recently turned its attention to reducing manufacturing costs, enacting an announced intention to remain with its products through later stages of the product life cycle. Intel is forthrightly competing in high volume markets. All of these are changes from the firm's earlier specialist perspective, although Intel continues to compete by means of very high research and development expenditures to support a "state-of-the-art" product line. Today the firm remains the acknowledged world leader in microprocessors. The shift, and attention to later-stage manufacturing and cost control, has enabled Intel to wring more profits from its discoveries, and thereby to fund further research.

Firms like Intel operate in an ongoing tension between stability and change, past commitments and future opportunities, freedom and direction. Theirs is not a simple competitive field. It is not enough to innovate; the end-game of highly precise manufacturing must also be mastered. It is not enough to master manufacturing of past products; the firms must also persist in innovation, in the face of constant competitive pressure from highly competent adversaries both in the United States and abroad. Old organization structures and old methods of management simply do not provide sufficient flexibility, sufficient vision or sufficiently robust and current understanding to cope with the challenges of change.

Motivating Innovation

To encourage and manage innovation, these firms maintain an on-going flow of innovations, foster a broadly shared criterion set for judging innovations, and explicitly permit deviations from plans. However, simultaneously they corroborate established strategic directions. These practices go well beyond standard theory descriptions of motivation, leadership or innovation because they include simultaneous emphasis on both stability and change, freedom and direction. These notable apparent contradictions are the very stuff of innovation, in high technology environments.

One contradiction resides in the idea flow itself. These companies require not merely technical ideas that can be specified for a particular individual decision, but rather a constant free flow of ideas from many sources, many perspectives and many individuals. Without such a free flow, these large firms will lack choices and options from which to continually refashion their strategic choices for the future. But, given such a flow of ideas, choices must be made. Even large companies cannot do all things for all people, nor pursue every good idea. This, in turn, means that some ideas will be denied or allowed to expire.

Motivation, in this context, is a dual problem: people must be induced to bring their ideas forth and share their insights, while at the same time they must be cushioned from the disappointment of having their ideas turned down. If denial is too constant, ideas will stop; if ideas stop, the innovation strategy will fail. Conversely, if no choice is made, too many options will be pursued, and all will starve from lack of resources—whether dollars, personnel, attention, manufacturing or marketing capacity. Absent success, innovators will correctly perceive that their ideas cannot be nurtured to acceptable maturity within the firm. They will take their ideas elsewhere, and so the innovation strategy will fail.

To achieve a successful strategy of innovation, then, we found three key features repeated across the companies we studied:

- They consciously work to maintain free flow of innovative ideas from many sources by encouraging new ideas.
- Managers open and share the strategic judgment process, depending on juniors for decisions and selection, not just for ideas. Because juniors also help select ideas, they, themselves help insure that selection criteria are fair, widely shared, and reasonable.
- Alternative channels are maintained for "legitimated subversion," to backstop even good current strategies against their inherent limitations.

Opening the Decision Process

To keep decision processes honest and to open them, realistic criteria are needed. The criteria must be broadly perceived as fair, reasonably objective, useful and typically nonpolitical. The criteria must also be broadly shared. *Marketplace success* is the criterion in these firms. Even very junior technical specialists are conversant with what constitutes market success, how it is measured and how their contributions can affect the firm's success or failure in the marketplace.

Beyond criteria, to keep contributors on board despite the necessity of choice, these firms also need second chances, "soft-landings" for failure, depending on its cause: you don't shoot the messenger, you don't penalize "any failure" with eternal damnation, as managers in these firms repeatedly noted: too rigorous a penalty will extirpate innovation quickly.

What is needed instead is a delicate balance—not letting people off the hook too easily, yet not stamping out risk-taking. Our sample firms accomplish this balance by getting people to internalize "the hook" in their own judgment. Some of these firms (and, of course, other highly innovative firms too) "recycle" people after failure, providing other opportunities to succeed.

Technological advances complicate matters further. "Technological failures" are a special case, and managing technological risk a special balancing act. Some failures are even desirable, the managers noted, since technological risk typically correlates with return. If a firm has no failures, its projects are likely too easy, and thus it is likely returning too low a profit. Obviously, those who "fail" because of this sort of risk are not to be blamed. Their personal career costs must not be too high. Equally, they must choose their own risks, exercising both judgment and courage in undertaking what *may* fail. Without genuine commitment and ownership, the possibilities of success will be subverted. Without some genuine risk, technological progress may be too slow.

Paradoxically, then, innovation is "controlled" only by opening up the process of strategic management and sharing control. Shared strategic management gives people genuine power. However, it is not grounded in altruism, so much as in pragmatism: no other method works. The practices of our high technology sample are very similar to Kanter's (1983) "empowerment" notions and akin to Quinn's (1980) incrementalism, and Wrapp's (1967) corridors of indifference. But the practices are specifically oriented toward innovation and the future of the firm, and specifically linked with extensive information systems to track operational performance, in the organizations we studied. However, the information systems are not used to "control" in the typical sense of that term. Instead, something far more like a cooperative collaboration is created, much along the lines described by Peter Drucker (1987, p. 2):

> We need to move from the traditional, military-style command structure to one that is more like a symphony orchestra. With the number of people the New York Philharmonic needs on stage to play Mahler's Seventh Symphony, traditional organizational structure would require deputy conductors, and section conductors to bring it off. But that's not the model for that organization. Instead, all of the players

report directly to the conductor. And the reason that's possible is because the conductor and the individual players all know the score.

Steering by Culture

The outcome, in these firms, is much informal guidance, shaped and evolved over time, and by means of deep acquaintance and familiarity among members. There are control systems, but they often serve as information channels, while control is internalized (Litterer, 1986). Control of innovation, an apparent contradiction in terms, seems to operate through organization culture. Culture fosters "innovation as a state of mind," resting on shared values and a shared vision of who we are, what is right, "how we do things." However, such practices create hazards, as well as benefits.

Pluses

Bearable control, feelings of potency and ownership, among organization members, flexibility to shift the firm into repeated innovations, effective teamwork, enthusiasm and investment.

Minuses

Over-intensity; potential for burnout, tunnel vision and "true believership," where paradigms govern what can be perceived; over-commitment to older views.

The paradoxical characteristics of culture pose a host of questions as yet unanswered by theory. Among them:

- how to manage a high-intensity, high-innovation change engine, without giving up control or responsibility, by sharing control and increasing responsibility;
- how to keep flow of innovation going, recognizing the complex interplay of systems, culture, leadership and strategy, individual investment and team loyalty;
- how to provide for necessary stability, while maintaining essential flexibility;
- how to share judgment and steering process;
- how to mind the process, and guide content, while opening content up (content comes mostly through others);
- how to avoid unduly heavy personal costs for members in high intensity, high involvement organizations; and
- how to do all of this simultaneously.

NEEDED ADVANCEMENTS IN THEORY AND PRACTICE

The nature of high technology and increasingly, the nature of competitive reality for most businesses, requires different management practices to deal with the changed demands and complexities. Old static, linear, rigid approaches simply cannot provide the flexibility and responsiveness necessary in a changeable, volatile, fluid competitive and technical environment. "Ad hoc" organizational practices also fall short, however: precision, control and meticulous management of a multitude of details is also required today—in the design process as in manufacturing, in marketing and in environmental scanning, in integrating functional activities and more. What we have been describing is clearly not "loose" management in any meaningful sense, although it is far different from the structured rigidities of past bureaucratic practice.

Current theory, and even current management vocabulary, are insufficient to comprehend the complexity managers increasingly face. Many older concepts are woefully impoverished both in usage and theory, and offer managers inadequate tools for reasoning about the challenges they face. We are not the only ones to observe these shortcomings, and other field-researchers have made contributions to moving management thinking beyond its older, traditional limitations. However, there is much yet to be done. Our purpose here is to emphasize these needs.

Technology

Although it is all too often seen as static, linear or unitary, technology changes almost constantly. This change is especially noteworthy in industries typically identified as "high technology." But an increasingly widening circle of basic and mature industries are also seeing new products, new processes and new materials impose massive change. The century-old automobile industry offers a case in point—although sheet metal, rubber and internal combustions are still the norm, GM and Volkswagen, Toyota, Ford and Volvo all use robotics, "engineered plastics" and much electronics. Change, under such conditions, is inescapable.

These changes are significant; they produce not only incremental improvements, but truly extraordinary advances. Technology changes, and with a fair degree of frequency changes in ways that make wholly new activities feasible or even mandatory. (For example, computer-controlled metal cutting relies on developments in computers and electronics, making possible degrees of precision in small, inexpensive machine tools simply inconceivable at any price, or obtainable at extremely high prices only, not long ago. As a result, closer tolerances at much lower prices can be expected as the norm—and those firms unable to provide them will lose out to more advanced competitors.)

Most organizations—and certainly those we have described here, including the "simplest" and "least complex," as classified in a single SIC code—are vitally concerned with multiple technologies at any instant. Such multiple technologies evolve or mature at different rates, and interact in nontrivial ways. A given product may draw upon several underlying production technologies, and multiple options for creating the product's desired characteristics may be available. Both the product and its evolution, and the processes and their evolution are significant sources for change.

In high technology firms particularly, managing a so-called "technology strategy" involves more than simply cranking out new products incorporating incremental improvements or enhancements, although that may also be required. Investments in R&D, particularly early stages investigations, constitute hedging—placing bets to insure options, not guarantees. Yet if such bets are not placed, the company may well forego participation in some new technology. Where the new technology is central, or provides some crucial advantage, the company may be compelled to purchase it from others (perhaps at an exorbitant price), or exit its chosen markets. Surprise developments by others, unanticipated by a firm blindly committed to existing technology for products and processes, can invite "technological mugging"—disastrous disadvantages in products or production capabilities, price or quality that can destroy market share.

Managing multiple technologies means managing multiple time-horizons, multiple risks and multiple development clocks: exploratory efforts are not necessarily predictable, although in contrast targeted development can be highly structured, with endpoints well specified. Both sorts of activity are necessary in highly volatile environments, requiring more complex and sophisticated thinking about technology and its management than many current models will support.

Structure

In our sample of high technology firms, structure is more a flexible tool for realignment than a rigid or confining skeleton. High technology firms use structure, anticipate that it will change and shift rather fluidly from one structural arrangement to another. The structure that is in place, however, does indeed specify quite explicitly who shall report to whom, who has responsibility for decisions, what a given entity's field of action is, and so on. These firms shift frequently from one definitive structure to another.

Unlike the structure of standard bureaucratic organizations, high technology structure is expected to change frequently. Consequently, there seems to be both a greater tolerance for structural explicitness, and higher expectation that structure shall match strategic intention and market need closely. If not, these managers seem to say, structure should be changed. By contrast, management theory has suggested that where structure fails to support de facto strategies and intentions, "informal structural arrangements" will take over instead. Informal structure may even subvert the espoused strategies and intentions of the firm, where these differ significantly from the strategies and intentions actually enacted or signaled by practice.

In high technology firms, there seems less of this subversion, and a closer linkage between intentions and actual practice, as well as between formal structure and informal mechanisms. Two reasons suggest themselves: the role of "informal" practices seems more supportive of strategy, and the frequent shifts of structure seem to improve alignment. What might have to be accomplished informally in a less change-tolerant, more bureaucratic organization can be in fact enacted in structure, where formal structure is a volatile changeable tool. The role of informal practices in our firms seems less a matter of pseudo structure, than of control and participation, to which we now will turn.

Control and Participation

Both control and participation are concepts inadequately defined and insufficient in explanatory power, in current theory, to explain the emerging practices of high technology firms. "Control" in the bureaucratic sense of restraining individual input or compelling unwilling compliance seems wide of the mark in a firm dependent on members' willing cooperation. Equally, while it is obvious that participation is desirable, to insure ownership and maintain engagement by key professionals in high technology firms, "participation" as this term is usually understood is simply not an adequate description of what happens. The participation we saw was neither the "consultative" sort that merely asks for advice, nor the delegative sort that completely cedes decisions to others. It was, instead, something far more complex. Terms like "self-control" (Litterer, 1986) and "shared strategic contribution" suggest the nature of the changes. Moreover, participation changes overtime, even for a given position, individual or level.

Strategy

Decisions and most particularly strategic decisions are impoverished by current theory in much the same fashion. Strategic decisions in our sample firms are both less definitively located, and far more pervasive, than current theory seems to suggest. Senior managers do not, for the most part, make single-point strategic decisions. Instead, strategy develops and emerges from multiple sources, with multiple inputs of multiple sorts. At bare minimum, "the strategic decision" is the finely grained result of many, many other activities (including information gathering and interpretation, analysis and insight that frame a context within which "a decision" can ultimately be taken). Even that schematic outline suggests that "participation" is far more than having a vote, or a voice, in "the decision." Similarly, strategic control is far more complicated than managers imposing criteria or monitoring results of subordinates. Instead, in the high technology firms we studied, control was widely shared and internalized as self-control, while strategic contribution included a wide array of information gathering, judgment and input activities.

Strategy, then, must be seen as far more than formulating strategy, and more than "controlling its implementation," in these firms. Instead, strategy management seems to center on creating and sustaining the processes, structures and systems that facilitate innovation. Such a process oriented focus emphasizes streams of information, repeated opportunities for broad input and self-correction in light of accurate, available information on results. This perspective rises above the traditional strategy formulation and implementation dichotomy to strategy process management and strategy process change, when needed.

"LOCAL KNOWLEDGE" AND CULTURE

What is "informal" has been often dismissed or deprecated in theory as contrary to, or apart from organizational goals, and thereby relegated to purely socio-emotional orientations. Such aspects take a far more central role in well-run high technology firms as a direct result of the organizational, managerial needs we have discussed. "Local knowledge" (Baba, 1988), grass roots insights and nonspecified assistance of nonmanagers is both the norm and the necessity in these firms. Much current theory assumes a fundamental split between the interests of managers and workers—while technical specialists are not quite considered in either camp. In high technology firms, such as those we studied, technical specialists are essential, and even workers often have crucial insights that must be taken into account. (Even older firms in more mature businesses like Ford Motor Co. have recently identified an important role for such local knowledge, as widely published accounts suggest [see, e.g., Mishne, 1988.]) In this context, we need more carefully delineated, redefined notions of how "formal" and "informal" activities might relate, and how they may legitimately be developed to reinforce one another and meet joint needs. Clearly, older assumptions that reinforce adversarial relations between "formal" and "informal," on management and workers, ill serve all sides. Our sample firms counted on informal contributions and local knowledge, and encouraged them.

Culture as a theoretical concept is likewise insufficiently defined to guide researchers in their investigation of the rich and compelling impact these firms' ways of operating have on their people. What is perhaps most striking is the degree of ownership and commitment, and thus the power, that culture has engendered in such firms as those we studied. Like "informal practices" and "local knowledge," organizational culture concerns both organizational agendas and individuals, in our sample firms. This culture is not a management tool, precisely; nor is it simply a counterbalance for rigid norms, a way to meet purely social needs, as we have argued. Instead, like informal organization, culture blends deeply into the firm's goals and strategic intentions and members' involvement. We need better ways to describe and understand culture than those we have; such approaches must address the often-unrecognized degree of essential ownership and self-investment so characteristic of high technology firms, both in negative and positive terms.

Teams

Teams and teamwork have long been subjects of investigation. A voluminous organizational development literature addresses techniques and methods of building teams, means of encouraging creative thinking, diagnostics, and the like. All too often, however, these literatures simply do not address the relationship of the team to other portions of the organization, nor to organizational goals. Strategic aims, so important to managers in practice, are frequently quite invisible in both OD literature and practice (Jelinek & Litterer, 1988). While team theory and team-development practices have much to offer in the management of innovation, these must be legitimately linked to strategy and to the larger organizational context, if they are truly to serve managers, organization members, and their needs.

Leadership

The role of the leader—in managing change and encouraging culture, opening the processes of strategy to wider involvement while also guiding and coordinating—vastly transcends traditional descriptions. The majority of the leader's tasks are not so much concerned with "decisions" as with process assurance, nuances and guidance. Impoverished theoretical concepts like charisma turn away from the hard work of understanding how leaders might infect followers with enthusiasm and commitment, by sharing the problems and goals of the organization, its guidance and its results.

CONCLUSIONS

Many theories and much of traditional management practice have grown up in a very different environment from what managers face today. In particular, the need for repeated innovation in the midst of constant competition has acquired great importance. Flexibility, fast response, cross-functional and interlevel integration, and adaptation to changing circumstances are all vastly more important now than in the past. So, too, are more complex, interconnected views of how people interact in organizations. Leadership and culture, strategy and participation, technology and structure and managerial stylistics are not independent variables. Instead, these elements interact complexly; so, too, should our theoretical perspectives; so, too, should our view of "best management practices."

While some managers have developed successful practices, these are less well-known than they might be. While theories have begun to recognize the increased complexity of contemporary management's tasks, there is much more to be done. It is time to bring our views of management up to date.

ACKNOWLEDGMENT

The conclusions presented in this paper are drawn from a forthcoming book, *The Innovation Marathon* by Mariann Jelinek and Claudia Bird Schoonhoven, (Basil Blackwell, 1990).

REFERENCES

Baba, M. (1988). *The local knowledge content of technology-based firms: Rethinking informal organization.* Paper presented at the Conference on Managing the High Technology Firm, University of Colorado at Boulder.

Burgelman, R.A., & Sayles, L.R. (1986). *Inside corporate innovation.* New York: The Free Press.

Burns, T., & Stalker, G.M. (1961). *The management of innovation.* London: Tavistock.

Daft, R. (1986). *Organization theory and design* (2nd ed.). St. Paul, MN: West Publishing.

Drucker, P.F. (1987). Productivity: Today's new meaning. *The Consultant Forum, 4*(3), 2-5.

Jelinek, M., & Litterer, J.A. (1988). Why OD must become strategic. In W.A. Pasmore & R.W. Woodman (Eds.), *Research in organizational change and development* (Vol. 2 pp. 135-162). Greenwich, CT: JAI Press.

Kanter, R.M. (1983). *The change masters: Innovation for productivity in the American corporation.* New York: Simon & Schuster.

Litterer, J.A. (1986). Elements of control in organizations. In M. Jelinek, J.A. Litterer, & R.E. Miles (Eds.), *Organizations by design* (2nd ed., pp. 447-457). Plano, TX: Business Publications.

March, J.G., & Simon, H.A. (1958). *Organizations.* New York: Wiley.

Mishne, P.P. (1988). A passion for perfection. *Manufacturing Engineering, 101*(4), 46-58.
Morrison, E.E. (1966). *Men, machines, and modern times.* Cambridge, MA: MIT Press.
Quinn, J.B (1980). *Strategies for change: Logical incrementalism.* Homewood, IL: Richard D. Irwin.
Schoonhoven, C.B. (1984). High technology firms: Where strategy really pays off. *The Columbia Journal of World Business, 19*(4), 5-16.
Schoonhoven, C.B. (1987). Combining strategies pays off fast, even in turbulent markets. *Planning Review,* 15(4), 36-41.
Wrapp, H.E. (1967). Good managers don't make policy decisions. *Harvard Business Review, 45*(5), 91-99.

THE TWIN DILEMMAS OF COMMITMENT AND COHERENCE POSED BY HIGH TECHNOLOGY

Janice M. Beyer

Both the general public and the academic community have become increasingly concerned about the implications high technology has for people and society now and in the future. One set of concerns is largely economic and pragmatic: How can we facilitate the development of more high technology so that our country, state or city will prosper and our citizens will have jobs in the future? Underlying these are less evident concerns derived from the images of high technology carried in popular books (Rogers & Larsen, 1984) and the popular press. They portray the business of high technology as very demanding—with high personal and financial risks, highly specialized and rapidly changing bodies of knowledge and expertise, unusual working conditions and fierce competition in unpredictable markets. Such demands pose many difficult challenges for managers, especially in managing the people whose efforts are crucial in developing new technology. One danger is that the way that individuals devise to cope with the demands of high technology in the short term may be damaging to organizations in the long term.

WHAT'S DIFFERENT ABOUT HIGH TECH FIRMS

The defining characteristics of high technology firms set them apart from other work organizations. As defined by the U.S. Bureau of Labor Statistics, high tech industries have twice the number of technical employees and double the R&D expenditures of other industries. The strong scientific/ technical basis of these industries drives the competition within them, creating markedly different market conditions for high tech firms than occur in other industries. In particular, because new technologies rapidly make existing technologies obsolete, the applications of new technologies can create or revolutionize markets and demands overnight (Shanklin & Ryans, 1987, p. 60). Under these conditions, success requires the adoption of a philosophy of "creative destruction," in which firms must destroy their own profitable existing technologies with something superior—for if they do not, someone else will (Shanklin & Ryans, 1987, p. 59).

In effect, change becomes the "order of the day" (Trebig, 1986, p. 3) and "the only constant" (Shanklin & Ryans, 1987, p. 4). Changing environmental demands are reflected in internal organizational dynamics. Observers of high technology firms have commented on the temporariness of much that goes on inside them (Kanter, 1987; Rogers & Larsen, 1984). As these firms continually devise new responses to the many changes they confront, internal boundaries shift around. Subunit identities, titles and roles change to keep up with technological developments, changing markets and products, new competitors and firm growth (Galbraith, 1985). Such changes have especially strong impacts in relatively small entrepreneurial firms like those in Silicon Valley, but even giant firms such as IBM cannot entirely escape the necessity to keep abreast of changing conditions with appropriate internal adjustments somewhere. Indeed, many analysts advocate trying to create entrepreneurial enclaves within large corporations, presumably to foster working conditions similar to those found in small, entrepreneurial firms. This paper speculates about some likely effects of continual change on those people who work in the changeful segments of high tech organizations.

In a recent talk summarizing her extensive observations of high technology firms, Kanter (1987, 1989) likened them to the croquet game in *Alice in Wonderland.* What made that croquet game memorable and pertinent was that nothing was fixed or predictable. The mallets were flamingos, who could move their heads just when a player tried to use them to strike a "ball." The balls were hedgehogs, who not only made a moving target, but could move around at will between plays without the players moving them. The wickets were live playing cards, which also moved around without players' consent. With the possibility that anything could move at any time, the traditional rules of the game did not work, and the game did not make sense to Alice. She found it very frustrating and did not want to continue playing once she discerned how it worked.

The extended metaphor of this unusual game is useful not only for dramatizing the very high rates of internal change Kanter observed in high tech firms, but because Alice's reactions point to some of the problems this continual change may pose for their members. Alice never really understood what the game was about; its lack of coherence disturbed her. Perhaps for this reason, she was not committed to playing it—in fact, she quit as soon as she could. Could unpredictability of events and the fluidity of boundaries of work roles, work groups, organizations, and even industries in high technology produce similar reactions? How many employees find the game of high tech disturbing once they experience it and decide not to continue to play? How do those who continue to play (1) develop and maintain work-related commitment under such conditions; and (2) create a coherent framework within which they can view what they are doing and what happens at work? The aim of this paper is to examine these questions and begin to sketch out some directions that research on them could pursue.

QUESTIONS RELATED TO COMMITMENT

In the organizational and management literatures, a well-established line of research has focused on various ways that individuals commit themselves to the organizations that employ them. Several conceptions of commitment have been advanced and received empirical support, suggesting that people become committed to their work settings in a variety of ways and for a variety of reasons. However, because high technology organizations are extreme or atypical in various respects—especially their fluidity—thinking about employees' commitment to them refocuses our attention on some basic, neglected questions: What conditions are necessary to elicit employees' commitment? What is the most likely basis of their commitment? To what entities will employees commit themselves in such work settings? With what consequences?

Will They?

The most obvious issue concerning commitment in high tech firms is whether people will commit themselves to anything at all if their work settings are very fluid and temporary. Commitment in organizations is usually defined in terms of a person's willingness to stay with an organization (Becker, 1960; Ritzer & Trice, 1969b), a psychological bonding to an organization akin to identification or loyalty (Buchanan, 1974), or a combination of the two (Mowday, Porter & Steers, 1982). In all of those senses, commitment takes time to develop. Research has repeatedly shown strong correlations between organizational tenure and organizational commitment (e.g., Mowday, Porter, & Steers, 1982, p.30; Ritzer & Trice, 1969a, 1969b; Stevens, Beyer, & Trice, 1978), suggesting that the commitment of members increases over time. Longitudinal research has confirmed this idea and has been tentatively interpreted as suggesting that commitment develops "through a self-reinforcing cycle. Commitment appears to influence other variables, which in turn, influence subsequent commitment" (Mowday, Porter & Steers, 1982, pp. 71-72).

Descriptions of high technology organizations suggest that many do not provide opportunities for such self-reinforcing cycles to occur. While the studies cited by Mowday, Porter, and Steers found strong effects of commitment after only three months, the employees studied had apparently not only remained in the same organization for that period to time, but also in the same job. But can commitment develop if people move around a lot inside organizations? Do different supervisors, tasks and coworkers provide the same opportunities for self-reinforcing cycles to accumulate as occurs with the same supervisors, tasks, and co-workers? Likely not, but descriptions of workers in Silicon Valley firms portray them as working long hours, and as having such "high personal commitment to entrepreneurial activity" that family life suffers (Rogers & Larsen, 1984, p. 153). Perhaps people in high tech firms can persist in essentially the same lines of activity, even though their role descriptions or work group boundaries change. Salancik (1977) argues that the extent of individuals' commitments to various behavior depends on the explicitness, revocability, volition and publicity of past related acts. If he is right, those work behaviors in high tech firms that are observed and unequivocal, irreversible, relatively unconstrained and public should produce more commitment than those that are not.

The type of commitment that involves staying with the same organization is another story. In Silicon Valley, "some estimate that job-changing may be 50 percent each year," with that among managers and engineers at "only" 30 percent (Rogers & Larsen, 1984, p. 87). Clearly, many high tech employees are not committed to staying with the same firm. Do the high rates of turnover reported give mobile employees enough time to become committed to their firms? Also, does the constant coming and going such turnover creates militate against those employees who do continue finding stable entities to which to commit themselves?

A related question is whether employees in organizations with such high turnover feel secure enough to commit themselves to their organizations (Morgan, 1988: 59). One executive reported that his firm dealt with this issue by committing itself to continued employment and retraining of its employees. Individuals were shifted around by giving them opportunities to serve 6 month stints on company-wide task forces that helped them to learn new skills and find new ways to employ them (Morgan, 1988, pp. 60-61). Such practices may work in firms where skills are not too intensive and specialized; they seem unlikely to work with highly specialized technicians or research scientists.

What Kind?

Implicit in the discussion thus far has been the possibility that individuals commit themselves in more than one way. Whether employees in high technology organizations are committed may depend on the basis of that commitment. While several conceptual and operational definitions of commitment have been used (see Angle & Perry, 1986; Mowday, Porter, & Steers, 1982, pp. 21-26 for reviews), only two bases of commitment have received much empirical investigation in the organizational and management literatures. The key distinction between them is whether the basis of the attachment is affective or calculative. A calculative attachment occurs when "bonding between member and social system results from a committed individual's linking the relationships to extrinsic outcomes on the basis of costs and benefits ...[an affective attachment] derives not from economic exchange, but from such processes as identification and internalization [Kelman, 1958]" (Angle & Perry, 1986, p. 33).

This distinction resembles Etzioni's (1975) categories of calculative and moral involvement, with moral involvement arising in response to the use of normative power and symbolic rewards, while calculative involvement arises in response to the use of remunerative power and economic rewards. His final form—alienative involvement—arises from the use of coercive power and has been relatively neglected in the study of work organizations. Because he assumed that the three forms of involvement form a single continuum of commitment, Etzioni implied that the emergence of the last two forms—calculative and alienative—would tend to undermine moral involvement, which he considered the strongest form of attachment. Management and organizational researchers have not pursued this insight nor linked commitment to the use of power as closely as Etzioni does .

New high technology organizations offer especially interesting sites in which to study the emergence of affective and calculative commitment and the trade-offs, if any, between the two. It would be interesting to assess the consequences if some firms rely more heavily on remunerative power based in economic rewards and incentives, while others develop strong normative power and affective/moral bonds. Also, do any firms present circumstances in which both forms of commitment flourish? Are the two forms or the conditions that give rise to them mutually reinforcing or inhibiting?

Good measures of these two forms of commitment exist; they are usually not included in the same study but could be. However, the measures of calculative commitment developed thus far tend to have more behavioral implications, while those for affective commitment tend to be more attitudinal (Mowday, Porter & Steers , 1982, pp. 24-26; Salancik, 1977; Staw, 1977). Ritzer and Trice (1969a, 1969b) developed a measure of calculative commitment based on Becker's (1960) side-bet theory, which basically equated commitment with staying in the social system to avoid losing accrued investments there. Items asked respondents how likely they are to leave the organization given various levels of several inducements. A number of other researchers have used their measures (Stevens, Beyer, & Trice, 1978) or variants of them (Alutto, Hrebiniak, & Alonso, 1973; Fukami & Larson, 1984). Porter and colleagues (Mowday, Porter, & Steers, 1982; Porter, Steers, Mowday, & Boulian, 1974) developed a widely used measure of affective commitment. It defines the committed individual as one who has a strong desire to remain a member, internalizes the values and goals of the organizations and is willing to work extra hard on its behalf. The first part of their definition overlaps somewhat with that in the calculative measure, but the items used are embedded in a list of rather affect-laden items.

While the two measures have usually produced results that agree—including those for tenure—they appear to be measuring quite different constructs, and not just stronger and weaker levels of the same phenomenon. Before Etzioni's ideas can be tested, it may be necessary to develop measures that are truly complementary rather than overlapping.

To What?

Past research has focused primarily on commitment to the employing organization, but there are clearly other possible targets for members' commitment, including family, and other entities in non-work life (Reichers, 1985, 1986). Only two have received much empirical attention: commitment to occupations and to unions. Other possibilities pertinent to high technology organizations include commitment to a work group, to an industry (Stevens, Beyer, & Trice, 1978, p. 394), to a product (nuclear power, personal computers or a particular drug), or to certain kinds of work behaviors (inventing or developing a new product).

Figure 1 gives a graphic illustration of how various targets of commitment could be reflected in engineers' staying or leaving their firms. At time t_1, Engineer 1 stays with Project A throughout its life cycle, suggesting he is committed to either the product or Organization 1. If he stays to begin a new project, Project B, he is probably committed to that organization. Engineer 2 leaves as the creative development work on Project A begins to decrease and goes to Organization 2, which is just beginning the

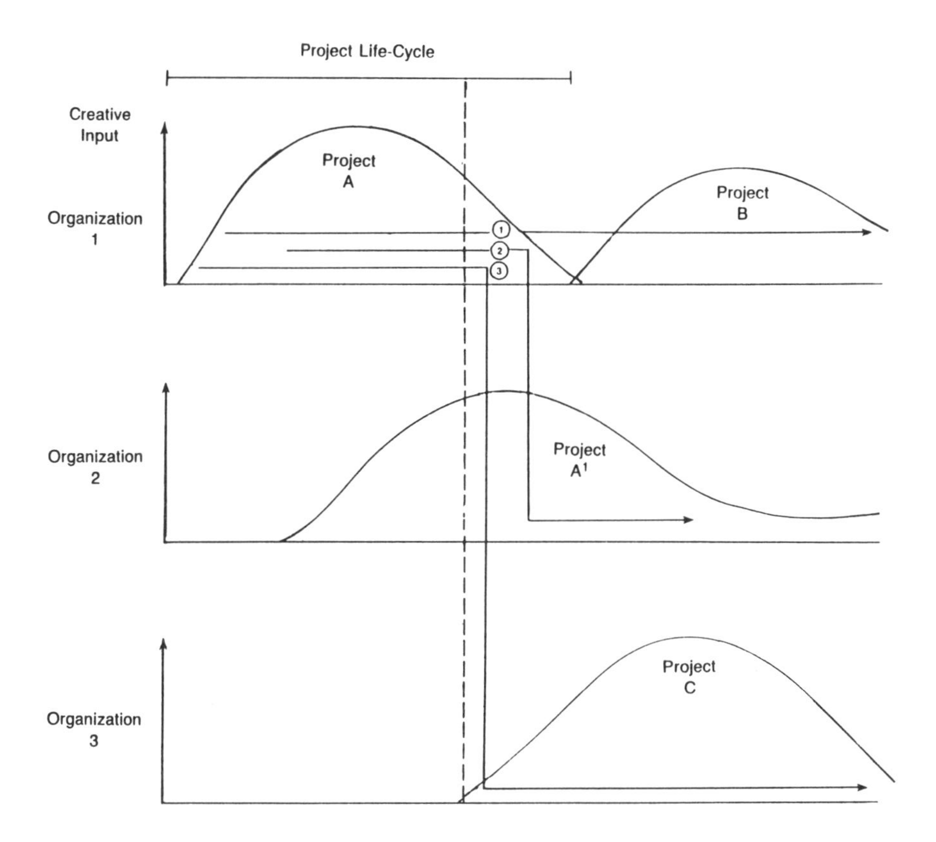

Note: The author thanks Michael Shumway for suggesting this illustration.

Figure 1. Project Life Cycles and Different Targets of Commitment among Engineers

development phase of a related project; this employee seems committed to developing a certain kind of product. Engineer 3 also leaves as the creative input in project A begins to decrease to join Organization 3, which is beginning Project C on an unrelated product; she appears to be committed to the creation process.[1] This illustration assumes other factors are equal; differences in the marketability of engineers' skills and other market considerations would, of course, complicate matters.

Such targets of commitment are highly speculative and have not been investigated. Another possibility, because of the fluidity of boundaries and memberships within high tech organizations, is that some occupational grouping becomes the target of commitment. Like the mobile cosmopolitan faculty in Gouldner's (1957) classic dichotomy, the most expert workers in these firms may identify with other persons doing the same sort of work and having the same kinds of experience, training and interests. Such a self-definition makes mobility between employers easier. It also facilitates banding together to try to control the conditions and content of work—a defining characteristic of true occupations and professions (Child & Fulk, 1982; Friedson, 1973). Sociologists have developed a substantial literature on occupations that might be helpful in considering how workers in high technology firms relate to their firms, their work, and each other (e.g., Ritzer, 1983; Rothman, 1987). The emergence and growth of occupational associations in high tech fields, for example, should not be ignored in trying to understand what happens inside firms.

Multiple Commitments

Once multiple entities are seen as appropriate targets for commitment, the same question arises as occurred for different types of commitment. Does having one target of commitment preclude or interfere with having other targets? Can people simultaneously have multiple commitments? Like all dichotomies, Gouldner's cosmopolitan/local distinction implied that people tend to have one or the other orientation, rather than both. Industrial relations scholars phrased this question in terms of whether workers could have dual loyalties to both employers and unions (Stagner & Rosen, 1965).

The first empirical study looking at dual loyalties focused on the relation between organizational and occupational commitment among personnel managers. Ritzer and Tricě (1969a, 1969b) found that the two forms of commitment were highly correlated, and thus apparently did not interfere with each other. Ritzer and Trice speculated that the dual loyalty they observed arose because the personnel occupations are relatively underdeveloped, lacking sufficient status or meaningful enough tasks to offer strong inducements to commitment. Assuming that people need to commit themselves to something in the work setting to make their work life meaningful and to find satisfaction in their work, they suggested that personnel managers psychologically hedge their bets, with their organizational commitment being a residual option exercised in the absence of strong inducements to occupational commitment (1969a, pp. 40-41).

These arguments have interesting implications for high technology firms, where many occupations are not well-established. In such a situation, members of high technology firms may also be disposed to hedge by committing themselves to their organizations when encouraged to do so. Given the long hours many high tech employees work (Kanter, 1987; Rogers & Larsen, 1984), it appears that these employees have committed themselves to something in the work setting; in the presence of conditions discouraging organizational commitment, they must develop other work-related attachments.

More recent studies of dual loyalty have looked at union and organizational commitment simultaneously (e.g., Angle & Perry, 1986; Fukami & Larson, 1984; Magenau, Martin, & Peterson, 1988). Although their methods vary somewhat, results substantially agree. Dual commitment is not unusual and most likely to occur in situations where union-management relations are good and job satisfaction is high. Only when union-management relations are bad or job satisfaction is low do employees apparently choose to commit themselves to only the union.

If these results are generalizable to other targets of commitment, they suggest that organizations that are able to accommodate themselves to professional and occupational goals will have more organizationally committed employees, while those that cannot or will not accommodate will force their employees to choose another target of commitment. Given the labor market demands in many high tech industries, employing organizations that force employees to choose may lose employees' loyalties and the employees themselves.

Recent research on technical systems in organizations also raises an interesting question about the likely targets of employee commitment in high technology firms. A recent study of manufacturing firms in New Jersey (Hull & Collins, 1987) found evidence for "Woodward's missing type"—technologies that are both small in scale and technically complex in terms of the technical expertise required of production workers. Flexible, computerized manufacturing systems would fit in their new category. While they had no data on the social conditions prevalent in these technologies, it seems feasible that their small scale could create social conditions similar to those found in the traditional craft technologies, which are small in scale but do not require complex knowledge. If high technology operations create, in effect, a technologically advanced version of the crafts, we should expect workers in such settings to develop strong loyalties toward occupational groups. For those high tech firms that use small-scale, technically complex technologies, the literature on the craft occupations may therefore offer useful insights (Rorabaugh, 1986).

Possible New Theoretical Directions

Craft occupations produce strong occupational groups, suggesting that one of the likely ways in which members of high technology firms will deal with various needs will be to form strong occupational subcultures (Trice & Sonnenstuhl, 1987) and occupational communities (Van Maanen & Barley, 1984). Since cultures not only bind people together, but provide ways of understanding their worlds (Beyer, 1981), this point will be addressed more fully in the discussion of coherence.

Another theme in the literature relevant to commitment and the changefulness of high technology organizations is what might be called "dual-track" organizations. A common thread running through many analyses of organizations (Ansoff & Brandenburg, 1971; Burns & Stalker, 1961; Mintzberg, 1979) is that firms need to be participative to foster creativity and innovation but relatively rigid and bureaucratic to achieve efficiency and peak competitiveness. More than a decade ago, Duncan (1976) proposed that organizations need to be "ambidextrous" in order to innovate successfully. He suggested devoting different parts of organizations to the creation of innovations and to their implementation. A variation of this theme was addressed by Galbraith (1985), who suggested that firms emphasize different types of structure at different points in their development. The problem with these analyses is that they pose difficult problems for managers without specifying exactly how to solve them. In particular, they fail to address how people become committed to living with the contrasts or drastic changes involved.

Fortunately, some empirical research suggests that the necessity or virtues of structural duality may be overstated. For example, wide participation in decision making was found to be positively related to the implementation of change in two studies (Beyer & Trice, 1978; Joyce, 1986), although Duncan theorized that successful implementation requires more centralized control. Apparently bureaucratic/mechanistic organizations are not always more efficient implementors than participative/organic ones. Perhaps their efficiency is purely theoretical. Why can organizations not be participative throughout to some degree and also be efficient? Japanese organizations seem to be.

Granovetter's (1973) classic paper of the strength of weak ties (1973) also raises a final, intriguing question about commitment in these fluid firms. In it, he shows how ostensibly weak ties can actually have strong effects. Perhaps strong ties such as commitment are not really necessary or desirable in high technology firms. Could people then use networks of weak ties for the same purposes? In their analysis of Silicon Valley, Rogers and Larsen (1984) emphasize the information exchanges that take place through the many close social networks there. A more recent study of a new venture breakfast club on Route 128 in the Boston area confirms that the weak ties formed during conscious networking activities had powerful consequences (Nohria, 1988). Many led to "new combinations"—new products, processes, firms, mergers, jobs, and so forth—within two months of the involved parties meeting at the club. Nohria likened the breakfast group to a bazaar, in which people judged one another's worth largely through their conversations and through reputation within a loose community of other specialists in the same area. Such networks could clearly provide bases for both calculative and affective commitment.

ISSUES RELATED TO COHERENCE

What Is Necessary?

If they indeed resemble the croquet game in *Alice in Wonderland,* high technology firms may provide natural settings in which to assess the strength of people's need for coherence and how they make sense, individually and collectively, of their work worlds as they encounter change and complexity.

What Is Possible?

High tech organizations have basically two options in adapting to their rapidly changing environments. They can either expect employees to change what they do, or they can change the mix of skills within their workforce through expansion or voluntary and involuntary turnover. Either way, many employees of high tech organizations will experience changes in their work roles—some in the same organization, others by changing employers. Entering a new organizational role or setting is "characterized by disorientation, foreignness, and a kind of sensory overload....There is no gradual exposure and no real way to confront the situation a little at a time. Rather, the newcomer's senses are simultaneously inundated with many unfamiliar cues" (Louis, 1980, p. 230). Entering a new role involves a variety of objective changes between the new and old work settings, particular features the newcomer experiences as contrasting with the old, and surprises about differences between individual anticipations and subsequent experiences. To cope with such novel circumstances, individuals cannot rely on preprogrammed cognitive scripts, but must engage in conscious thinking (Abelson, 1976; Langer, 1978).

Since some employees of high tech organizations have highly specialized skills, they presumably have well-developed cognitive scripts and routines for carrying out their work. The changing markets, technologies and products of high tech firms, however, make it unlikely that these scripts fit new roles or organizations so well that surprises or contrasts do not occur. In any case, there will be some objective differences from the old to the new setting, necessitating some leave taking from old roles and some letting go of old practices (Louis, 1980, p. 231).

Also, because of the fast pace of external and internal changes, high technology organizations may suddenly find that past organizational routines for developing understandings are ineffective. Past sources of information may be irrelevant, so new scanning routines need to be developed. Interpretive schemes used in the past may not fit present circumstances. Thus the patterns of behavior by which learning and understanding (Daft & Weick, 1984) occurred in the past may need to be radically altered to apprehend the present. In the face of widespread uncertainty, high tech organizations may need to develop ways to tolerate and retain diverse ways of seeking coherence, as well as diverse understandings. Such devices as loose-coupling (Weick, 1976), good faith and overlooking (Meyer & Rowan, 1977) may be useful in coping with valuable internal contradictions and conflicts. Such loose connections between elements, however, only create more ambiguity (Weick, 1985).

One way of looking at organizations, that seems especially pertinent to high tech firms, is as scientific communities with self-interests in which members interact to construct knowledge about themselves and their environments (Weick, 1979). Research on decision making in universities (Pfeffer, Salencik, & Leblebici, 1976) indicates that self-interests are more powerful in conditions of uncertainty than in conditions of relative certainty. Given their uncertainty, do high tech organizations give rise to especially powerful self-interests? How are these self-interests reconciled to arrive at coherent maps and explanations of a changeful world? How do members of these firms deal with the conflicts provoked by their self-interests?

Where Does It Come From?

When faced with uncertainties, people can look to various entities for coherence, as well as for commitment. It seems reasonable to suppose that the sources of coherence and the targets for commitment might be the same. Cultural views of organizations, in particular, might expect them to be the same.

The issue then becomes, as it was for commitment, to determine what role the organization plays and what role other social entities play in providing coherence for members. Again the occupation seems the leading alternative to the organization. The question becomes: Do people in high tech firms look primarily to their organizations or to their occupations to provide a coherent view of their work world? Under what conditions is one source preferred over the other? Is one source universally preferred, or is the choice of a source more a contingent or an individual matter? In general, how much do people rely on others to help formulate a coherent picture of the world? To whom in the organization do they turn? Recent theoretical work on cognition in organizations (Daft & Weick, 1984; Sandelands & Stablein, 1987) may provide a useful starting point to explore such issues in actual work settings.

How Shared?

There are two questions underlying this phrase. One asks about the degree to which understandings that provide coherence to one person are shared by others. The second asks by which social or other processes these understandings come to be shared.

Most organizations attempt, in one way or another, to socialize their members. Van Maanen and Schein (1979) identified six polar dimensions by which the processes that go on during socialization can be categorized. Using these dimensions, Jones (1986) found that how people were socialized produced different attitudinal outcomes. While the social dimensions of socialization and the content of information were relatively important in reducing uncertainties surrounding the entry process, both collective and formal socialization were less so (Jones, 1986, p. 275). While these findings suggest that informal socialization is very important, it may work differently in high tech firms than it does in less fluid settings. Have co-workers been around long enough to help newcomers learn the ropes? Could some degree of coherence be produced effectively through formal training or orientation, especially if it were geared to indoctrinating newcomers into what makes this firm, product, or process different?

It would obviously be an oversimplification to assume that all socialization takes place within prespecified processes. Many of the understandings that people share emerge gradually from the informal give and take of social life. In his classic analysis of the human group, Homans (1950) pointed out that people who interact frequently are likely to begin to share sentiments, especially if they are similar socially in other respects, like age, social class or education. These conditions are met in high tech organizations, where the unstructured nature of many tasks requires extensive and intense social interaction (Kanter, 1987). Homans predicted that such social processes would produce cohesive informal groups. They also provide a good basis for subculture formation if two additional conditions are met. First, the social interaction persists over a long enough period of time for the understandings to become tacit—for what are at first self-conscious ideas to become unconscious and taken-for-granted. The longer the interaction, the more powerful and tacit such ideas are likely to become. Second, ideas are not only genuinely shared but unique to a particular identifiable group. The most common way for such shared understandings to arise is in response to shared experiences, especially difficult and puzzling ones. Again, the uncertainties and complexities inherent in many tasks in high tech organizations may help to create such unique and shared understandings. An example may be what Baba (1988) calls "local knowledge."

Possible New Theoretical Directions

This brief analysis suggests that high tech organizations provide especially fertile ground for the growth of subcultures, especially subcultures based on occupational distinctions within these firms. If what Nohria (1988) observed in the Boston area is true elsewhere, at least some of the people who work in high tech firms tend to remain within a single geographic area as they move in and out of specific firms and jobs. In order to find jobs and to gain a reputation within their highly specialized areas of expertise, such workers must find others like themselves with whom they can share their expertise and learn about new developments. Especially for those who come to share more than work-related interests, such relative geographical stability combined with job mobility should form a structural basis for genuine occupational communities to form.

The signs of this development include ethnocentrism, using the occupational group as a reference group, sharing unusual emotional demands, developing a favorable group image and identity and the extension of the activities of the occupational group into nonwork life (Trice & Sonnenstuhl, 1987; Trice & Beyer, 1990). All of these signs were evident in Rogers and Larsen's (1984) fascinating account of Silicon Valley.

The second theoretical stream that may be relevant to high tech organizations is the study of ideologies and scientific paradigms. Briefly, the empirical findings on the effects of scientific paradigms in university governance show that when members of fields lack consensus over what they believe in, there is more conflict and difficulty in reaching decisions (Lodahl & Gordon, 1972), and that the use of power is more likely to decide such things as resource allocations (Pfeffer, Salancik, & Leblebici, 1976).

The third stream comes from the research on cognitive consistency. Balance theories like Heider's (1958) suggest that if two elements that are both valued contradict each other or are seen as in opposition, the tension created by this lack of balance will be resolved by rejecting one of the conflicting elements and accepting the other. For example, if people find they are in disagreement with a person they like over an idea that is important to them, they will either discard the idea or begin to like the person less.

If the highly specialized employees in high tech firms have unique and absorbing experiences that yield unique understandings, will they be able to maintain cordial relations with those who hold other views of the world? Or will they tend to develop antipathies to others who do not share their unique views? Universities are notorious for strong antipathies between faculty who hold diverse professional views of the world. To the degree that high tech firms develop equal levels of specialized expertise and intensely involving experiences in which it is exercised, they may also be creating conditions ripe for endemic conflict. This raises the challenging question of how these firms can simultaneously cultivate diversity, commitment *and* good will.

POSSIBLE DANGERS AND PROBLEMS

This paper has focused on dilemmas posed by some of the characteristics of high technology firms. Only two have been explored: the changefulness of these organizations and their high degree of specialization. Other features of these firms would probably lead to an equally long list of questions. However, it was not the intent of this paper either to explore any of these issues in detail, or to provide a comprehensive list of relevant questions and issues. Rather, its purpose is to provoke interest and some thought about some issues, and then suggest some ways in which they might be tackled.

Rather than summarize all that has gone before, it may be more useful to distill some of the main dangers and problems for these firms suggested by this analysis. The first of these is low organizational commitment and its likely consequences. One consistent consequence, well-documented in the literature, is high turnover (Mowday, Porter, & Steers, 1982). The most recent comprehensive analysis found there was a direct relation between commitment and intention to leave, which in turn is the best predictor of actually leaving (Lee & Mowday, 1987). But, as various analysts have pointed out, all turnover is not bad. Some of the turnover that occurs in high technology firms could be desirable, if the firm no longer needs the particular skills those individuals possess. Other turnover could be beneficial if the individuals involved simply don't fit into the culture or lack needed technical skills. The prime danger, of course, is that the conditions producing low commitment may lead to those employees the firm wants to retain going while those the firm does not need as much stay. These considerations suggest that managers adapt a targeted approach to trying to build commitment.

The second major problem for these firms could well be the familiar problems of integration first highlighted by Lawrence and Lorsch (1967). Like functional organizations, these firms may breed enclaves of specialists with myopic views of the world. This would be especially true if strong occupational subcultures were to develop within them. High tech managers will need to devote time and energy to integrating diversity and to dealing constructively with the inevitable conflicts and turf wars that emerge from so many pronounced differences. They will also have to choose whether to rely more on structural or cultural mechanisms for needed integration.

The final danger for these firms and for society is that they will be dangerous for the health of their employees. With little fixed structure, much ambiguity and change and strong possibilities for endemic conflict, it is hard to see how people will cope with all of the tensions. As the organizations and their members get older, will everyone be able to bear up under the accumulated tensions? Or will high technology firms breed sick people and sick families? Some of the firms seem to have developed rather macho kind of cultures, in which no one admits that anything is too much. Does this kind of culture provide needed social supports when people do not succeed or fail? It is not clear. If they fail, is this good for their members or for their society?

The question for now is whether we can have firms that are innovative, competitive *and* preserve our wider social values. The answers we come up with will help to create the social opportunities and problems of the next century.

ACKNOWLEDGMENT

The author would like to thank George Huber, Michael Shumway, Sim Sitkin, Ray Smilor, and Harrison Trice for advice and comments on this paper.

REFERENCES

Abelson, R.P. (1976). Script processing in attitude formation and decision making. In J.S. Carroll & J.W. Payne (Eds.), *Cognition and social behavior* (pp. 33-46). Hillsdale, NJ: Laurence Erlbaum.

Alutto, J.A., Hrebiniak, L.G., & Alonso, R.C. (1973). On operationalizing the concept of commitment. *Social Forces, 51*, 448-458.

Angle, H., & Perry, J.S. (1986). Dual commitment and labor-management relationship climates. *Academy of Management Journal, 29*, 31-50.

Ansoff, H.I., & Brandenburg, R.G. (1971). A language for organization design: Part II. *Management Science, 17*, B717-B731.

Baba, M.L. (1988). *Managing local knowledge to improve performance in high technology organizations.* Paper presented at the conference, Managing the High Technology Firm, University of Colorado at Boulder.

Becker, H.S. (1960). Notes on the concept of commitment. *American Journal of Sociology, 66*, 32-40.

Beyer, J.M. (1981). Ideologies, values, and decision making in organizations. In P.C. Nystrom & W.H. Starbuck (Eds.), *Handbook of organizational design* (Vol. 2, pp. 166-202). London: Oxford University Press.

Beyer, J.M., & Trice, H.M. (1978). *Implementing change: alcoholism policies in work organizations.* New York: Free Press.

Buchanan, B., II. (1974). Building organizational commitment: The socialization of managers in work organizations. *Administrative Science Quarterly, 19*, 533-546.

Burns, T., & Stalker, G.M. (1961). *The management of innovation.* London: Tavistock.

Child, J., & Fulk, J. (1982). *Maintenance of occupational control. Work and Occupations, 9*, 155-192.

Daft, R.L., & Weick, K.E. (1984). Toward a model of organizations as interpretation systems. *Academy of Management Review, 9*, 284-295.

Duncan, R.B. (1976). The ambidextrous organization: Designing dual structures for innovation. In R.H. Kilmann, L.R. Pondy, & D.P. Slevin (Eds.), *The management of organization design* (pp. 167-188). New York: North-Holland.

Etzioni, A. (1975). *A comparative analysis of complex organizations* (rev. & enl. ed.). New York: Free Press.

Friedson, E. (1973). Professions and the occupational principle. In E. Friedson (Ed.), *Professions and their prospects* (pp. 19-33.) Beverly Hills, CA: Sage.

Fukami, C.V., & Larson, E. (1984). Commitment to company and union: Parallel models. *Journal of Applied Psychology, 69*, 367-371.

Galbraith, J.R. (1985). Evolution without revolution: Sequent computer systems. *Human Resource Management, 24*, 9-24.

Gouldner, A.W. (1957). Cosmopolitans and locals: Toward an analysis of latent social roles. *Administrative Science Quarterly, 2*, 281-306.

Granovetter, M.S. (1973). The strength of weak ties. *American Journal of Sociology, 78*, 1360-1380.

Heider, F. (1958). *The psychology of interpersonal relations.* New York: Wiley.

Homans, G.C. (1950). *The human group.* New York: Harcourt, Brace.

Hull, F.M., & Collins, P.D. (1987). High-technology batch production systems: Woodward's missing type. *Academy of Management Journal, 30*, 786-797.

Jones, G.R. (1986). Socialization tactics, self-efficacy, and newcomers' adjustments to organizations. *Academy of Management Journal, 29*, 262-279.

Joyce, W.F. (1986). Matrix organizations: A social experiment. *Academy of Management Journal, 29*, 536-561.

Kanter, R.M. (1987, August). *Men and women of the Change Master Corporation (1977-1987 and beyond): Dilemmas and consequences of innovations in organization structure.* Distinguished Address, Academy of Management Meetings, New Orleans, LA.

_______ (Forthcoming). *The great corporate balancing act: Managing the fallout from the entrepreneurial explosion* (tentative title). New York: Simon & Schuster.

Kelman, H.C. (1958). Compliance, identification, and internalization: Three processes of attitude change. *Journal of Conflict Resolution, 2*, 51-60.

Langer, E.J. (1978). Rethinking the role of thought in social interactions. In J.H. Harvey, W. Ickes, Y R. F. Kidd (Eds.), *New directions in attribution research* (Vol. 2, pp. 35-38). Hillsdale, NJ: Lawrence Erlbaum.

Lee, T.W., & Mowday, R.T. (1987). Voluntarily leaving an organization: An empirical investigation of Steers and Mowday's model of turnover. *Academy of Management Journal, 30*, 721-743.

Lawrence, P.R., & Lorsch, J.W. (1967). *Organization and environment.* Boston: Harvard University Press.

Lodahl, J., Beyer, J.M., & Gordon, G. (1972). The structure of scientific fields. *American Sociological Review, 37*, 57-72.

Louis, M.R. (1980). Surprise and sense making: What newcomers experience in entering unfamiliar organizational settings. *Administrative Science Quarterly, 25*, 226-251.

Magenau, J.W., Martin, J.E., & Peterson, M.M. (1988). Dual and unilateral commitment among stewards and rank-and-file union members. *Academy of Management Journal, 31*, 359-376.

Meyer, J.W., & Rowan, B. (1977). Institutionalized organizations: Formal structure as myth and ceremony. *American Journal of Sociology, 83*, 340-363.

Mintzberg, H. (1979). *The structuring of organizations.* Englewood Cliffs, NJ: Prentice-Hall.

Morgan, G. (1988). *Riding the waves of change: Developing managerial competencies for a turbulent world.* San Francisco: Jossey-Bass.

Mowday, R.T., Porter, L.W., & Steers, R.M. (1982). *Employee-organizational linkages: The psychology of commitment, absenteeism, and turnover.* New York: Academic Press.

Nohria, N. (1988, January). *Organizing to identify and link variety-generating expertise.* Paper presented for Management Area, Graduate School of Business, New York University.

Pfeffer, J., Salancik, G.R., & Leblebici, H. (1976). The effect of uncertainty of the use of social influence in organizational decision making. *Administrative Science Quarterly, 21,* 227-245.

Porter, L.W., Steers, R.M., Mowday, R.M., & Boulian, P.V. (1974). Organizational commitment, job satisfaction, and turnover among psychiatric technicians. *Journal of Applied Psychology, 59,* 603-609.

Reichers, A. (1985). A review and reconceptualization of organizational commitment. *Academy of Management Review, 10,* 465-476.

________ (1986). Conflict and organizational commitments. *Journal of Applied Psychology, 71,* 508-514.

Ritzer, G. (1983). *Working: Conflict and change* (3rd ed.). Englewood Cliffs, NJ: Prentice-Hall.

Ritzer, G., & Trice, H. M. (1969a). *An occupation in conflict: A study of the personnel manager.* Ithaca, NY: New York State School of Industrial and Labor Relations, Cornell University.

________(1969b). An empirical study of Howard Becker's side-bet theory. *Social Forces, 47,* 475-479.

Rogers, E.M., & Larsen, J.K. (1984). *Silicon valley fever: Growth of high-technology culture.* New York: Basic Books.

Rorabaugh, W.J. (1986). *The craft apprentice: From Franklin to the machine age in America.* New York: Oxford University Press.

Rothman, R.A. (1987). *Working: sociological perspectives.* Englewood Cliffs, NJ: Prentice-Hall.

Salancik, G.R. (1977). Commitment and the control of organizational behavior and belief. In B.M. Staw & G.R. Salancik (Eds.), *New directions in organizational behavior* (pp. 1-54). Chicago: St. Clair Press.

Sandelands, L.E., & Stablein, R.E. (1987). The concept of organization mind. *Research in the Sociology of Organizations, 5,* 135-161.

Shanklin, W.L., & Ryans, J.R., Jr. (1987). *Essentials of marketing high technology.* Lexington, MA: Lexington Books.

Stagner, R., & Rosen, H. (1965). *Psychology of union-management relations.* Belmont, CA: Brooks-Cole.

Stevens, J.M., Beyer, J.M., & Trice, H.M. (1978). Assessing personal, role, and organizational predictors of managerial commitment. *Academy of Management Journal, 21,* 380-396.

Trebig, J.G. (1986). The take-off company: Self management and flexible structure. In R.W. Smilor & R.L. Kuhn (Eds.), *Managing take-off in fast-growth companies* (pp. 3-18). New York: Praeger.

Trice, H.M., & Beyer, J.M. (1990). *The cultures of organizations.* Book manuscript in progress.

Trice, H.M., & Sonnenstuhl, W.J. (1987). *Linking organizational and occupational theory through the concept of culture.* Paper presented at the American Sociological Association meetings, Chicago.

Van Maanen, J., & Barley, S.R. (1984). Occupational communities: Culture and control in organizations. *Research in Organizational Behavior, 6,* 287-365.

Van Maanen, J., & Schein, E.H. (1979). Toward a theory of organizational socialization. *Research in Organizational Behavior, 1,* 209-264.

Weick, K.E. (1976). Educational organizations as loosely coupled systems. *Administrative Science Quarterly, 21,* 1-19.

________(1979). Cognitive processes in organizations. *Research in Organizational Behavior, 1,* 41-74.

________(1985). Sources of order in underorganized systems: Themes in recent organizational theory. In Y.S. Lincoln (Ed.), *Organizational theory and inquiry: The paradigm revolution:* 106-136. Beverly Hills: Sage.

MANAGING THE CULTURE OF INNOVATION:

THE SYNTHESIS OF MULTIPLE DIALECTICS

Gerald E. Fryxell

Environmental turbulence, technological discontinuities, and aggressive international competition dictate that industrial R&D must be given a more central role on the corporate stage (e.g., Ansoff, 1985; Foster, 1986). Yet, in spite of the apparent consensus among academicians and practitioners about the centrality of innovation, the performance of industrial R&D units continues to fall short of aspirations (*"Competiveness Survey,"* Review, 1987). In spite of the temporary economic boost from our devalued currency, it is becoming increasingly evident that solutions to the real lag in competitiveness will be more elusive than reversing misplaced emphasis. Renewed competitiveness, it appears, must be preconditioned by a more fundamental change involving our most basic premises and assumptions about how to manage innovation.

The purpose of this paper is to argue that innovative cultures are inherently dialectical and to illuminate the importance and function of culture in the effective management of imbedded innovative groups. For clarity the term "imbedded innovative group" is used to refer to an identifiable group working toward the creation and application of new ideas within a larger and less-innovative corporate context. This is meant to distinguish between innovative corporate units, which is the focus here, from entrepreneurial contexts. This will be accomplished in four parts. The first section clarifies terminology and identifies some of the inherent problems in managing an innovative culture. The second section integrates the contributions of three theoretical perspectives of organization culture. The relationship among specific norms and values are compared within a framework of cultural form and function. In the third section of this framework inherent contradictions in the cultures of imbedded innovative units are identified and their dimensions are discussed. Finally, the implications of this dialectic conceptualization of innovative cultures for management and management education and training are discussed.

CONCEPTUALIZATIONS OF CULTURE

"Culture" and "ideology" are very ambiguous terms in the language of modern management. This is evidenced by the multiple definitions of culture that have been proposed from rather disparate conceptual vantage points (Lucas, 1987; Smircich, 1983). In this paper, innovative unit culture is defined as shared cognitions in the form of values, beliefs and assumptions that are common in imbedded innovative organizational departments and that are instrumental to the accomplishment of organizational goals. Several aspects of this definition warrant further clarification.

First, an innovative department is a formally prescribed group of organizational members whose purpose is either to introduce change, produce new ideas or to otherwise "flesh out" a strategy that may be ill-defined because of environmental uncertainty. One characteristic of such a group is that their activities can only be prescribed within very broad parameters, yet their actions and decisions impinge upon most other departments. An R&D unit, a particularly creative advertising group, or a strategic management team would be examples of such a group. On the other hand, a testing laboratory or an engineering department would not be innovative departments for the purposes of this discussion because their activities may be systematically prescribed.

Second, the intention is to focus on shared cognitions that will be characteristic of effective imbedded innovative units. Thus, the unit of analysis is the department, although a relationship to portions of an overall "corporate culture" is important. Thus, it is important to maintain the proper level of analysis.

Third, this definition includes those shared cognitions that are instrumental (in the sense that they contribute in some way to the realization of corporate goals as formulated by a top management group) and "manageable" (in the sense that interventions can alter or strengthen them). The result is a rather circumscribed notion of culture; however, the intention is to enhance the clarity of the discussion and increase its managerial relevance.

Finally, the term "ideology" is intended to refer to more fundamental premises that are more broadly embraced and less consciously evoked. This is evident in Lodge and Vogel's (1987, p. 3) characterization of ideology as "timeless, universal, noncontroversial notions." For example, premises regarding specialization, human nature and economic efficiency are exemplary of managerial ideologies. Overall, this terminology permits a distinction between corporate culture as a more shallow (i.e., more consciously accessible values and norms rather than unevoked premises or assumptions), unique (i.e., either by virtue of its corporate membership or its professional affiliation) and manageable phenomenon than ideology.

Problems of Prescription

Due largely to theoretical grounding in Burns and Stalker's (1961) notion about "organic structures," conceptualizations of guiding or managing innovative processes take on a vague inconsistency. There has been a strong inclination to equate all forms of innovative effort with freedom. Along the way the notion of "controlled innovation" has become a type of managerial oxymoron. However, if innovation is to avoid merely increasing a firm's inventories of new ideas, it must be channeled and integrated. Unfortunately, this organic conceptualization has promoted its own oxymoron—"uncontrolled organization"—and has limited the development of useful prescriptions for the bewildered practitioner. Currently, the literature offers little more than a few universalistic prescriptions.

It has been proposed, for example, that the culture of innovation should be "supportive of failure" (Peters & Waterman, 1982). Similarly, Pinchot (1985, p. 199) colorfully proposes 10 "hills to die on," which include "tolerance of risk, failure, and mistakes." While these admonitions certainly harbor nuggets of truth, should not an equally strong case be made for the support of success, evaluation and direction? Is it possible that these equally relevant reminders have been shorted merely to avoid betraying a Janus-like equivocality in our knowledge?

With respect to the management of innovative cultures, it is likely that universal prescriptions are as much a reflection of our ignorance rather than our knowledge of the phenomenon. In time, an understanding of important contingencies may permit more specific and relevant suggestions for the managers of these units. However, this tendency in prescription to emphasize only half of dialectical reality (e.g., such as the requirements of both freedom and control in the management of innovation) is, at least partially, ideological.

Ideology and the Management of Innovation

An additional impediment to managing innovative cultures is the limited appreciation of the interaction of ideology and culture. Outmoded assumptions, premises and values about innovation are unusually tenacious and debilitating, ironically because of the high value society places on innovation. Consider the emotional attachment to notions of: "Yankee ingenuity"; the worn out myth that the Japanese are "organizational clones" who merely imitate our accomplishments; or the trauma to the national psyche subsequent from the "Challenger disaster." The effect of the tenacity of ethnocentric bias and rationalizations embracing historical success retard needed ideological adaptation and undermine global competitiveness (Lodge & Vogel, 1987). Lethargic educational institutions further contribute to this mismatch between outmoded ideological assumptions and competitive global realities. This poses a serious competitive disadvantage, even for progressive firms, as they must incur additional expenses of organizational development.

There appears to be similar ideological mismatching in our domestic industrial establishment. Environmental realities have changed in ways that make the prevailing romanticism associated with Edison's prototype labs and individual-based invention an inaccurate model of industrial innovation. The vast bulk of industrial R&D does not seek novel and discontinuous scientific knowledge, but rather the incremental and methodical application of the previously known. This is not to argue that discontinuous invention isn't important, but to call attention to the disparity between practice and premise. Even the discontinuities, when they occur, must be exploited in this way. This disparity between practice and premise is also underscored by the increasing incidence of joint ventures, heightened concern with technological espionage and the inapplicability of traditional legal mechanisms to this aspect of the corporate relations (Miller, 1987).

As indicated by a recent Harvard business review survey ("Competitiveness Survey," 1987), global competitive realities require new ideological premises based on holism and integration. This is underscored by the critical need for "downstream coupling" (Ansoff, 1985) of R&D with engineering, production, marketing and distribution. This will require a broader, more strategically-oriented R&D unit as implementation and integration become more salient and the composition of the modern imbedded innovative unit shifts from being an aggregation of inventors to a team of creative members with strong linkages to other units. Furthermore, this ideology needs to be based on an appreciation for synthesis—for an appreciation and reconciliation of imbedded innovative units as "cultures of contradiction."

Given this incongruity of the "ideology of innovation" with its "reality," the domestic institutions of industrial innovation (i.e., as purveyors of ideology, which are slow to adopt new premises) undoubtedly frustrate any managerial interventions to change it. As a result, they tend to reproduce technically sophisticated R&D managers who have been armed with powerful frameworks for understanding physical reality that bears little correspondence to the socio-economic reality they must manage.

In summary, there is a substantial gap between the problems of managing the R&D unit and the illumination provided by the literature. This is due not only to the complexity of the phenomena and its external linkages, but to a mismatch between managerial preparation and its social reality (Ellul, 1964). In the next section the dialectical nature of this social reality is explored by analyzing the shared cognitions of its culture from three disparate conceptual vantage points. This will facilitate an understanding of the functions that innovative cultures may perform and will bring its contradictions into sharper relief.

A CULTURE OF CONTRADICTIONS

Three intellectual streams of literature are useful for understanding this circumscribed conceptualization of imbedded innovative cultures: functionalism, utilitarianism (i.e., transaction costs analysis), and conflict theory.[1] These three perspectives are contrasted in Table 1. Across the top, each perspective is listed with a summary of the main contributory "themes." Along the vertical dimension, the characteristic forms of cultural cognitions are distinguished based on Werner & Schoepfle's (1979) classification of all cultural cognitions into one of three forms: (1) heuristic guides to behavior and decisions; (2) cognitions regarding cause and effect relationships; and, (3) descriptive knowledge about the characteristics of objects or meaning attached to symbols. These two dimensions—contribution and form—are combined to form a matrix of cells where specific examples of imbedded innovative cultural content may be considered.

Functionalism

Functionalist views of innovative cultures share a predisposition to view shared cognitions as contributing toward the well-being of the unit and, usually, the entire organization (Lucas, 1986). These contributions about innovative cultures have drawn heavily from systems theory and thus tend to impart an organismic-like quality to the interdependence of its departments. As a result, culture in innovative units is important in that

Table 1. Cultural Cognitions of Imbedded Innovative Units

Theoretical Perspective	*Functionalism*	*Utilitarianism Transaction cost analysis*	*Conflict Theory*
Contribution	Coping with technological complexity and uncertainty; Autonomous strategic behavior; Channeling collective commitment; Restraining egoism	Economic efficiency; Coping with transaction uncertainties	Legitimation of distribution criteria; Accommodation of value cleavages
Form	Exemplary content		
Guides to Decisions (Values & Norms)	Keeping up w/ field & publication; Sophistication over pragmatism; Value of challenging strategists' assumptions & "bootleging"; Collectivism & wholism; Disregard of formal procedures	Belief of collective effort as consistent with self interest; Values of publishing	Mission statements; Legitimacy of criteria for distribution of benefits; Professional ethics; Appropriateness of managerial intervention
Cause/Effect Knowledge (Premises & Assumptions)	Time orientation; Scientific theories; Knowledge of environmental change processes;	Knowledge of transaction outcomes; Belief in long term justice—knowledge of opportunitistic potential in transaction relationship	Values emphasizing altruism over material benefit; Knowledge linking top management to overall firm performance
Descriptive Knowledge (Facts & Characteristics)	Requisite knowledge of environmental phenomena; Jargon & unit symbols; Knowledge of other, cultural distinct units; Totemic corporate symbols and ceremony (Collective consciousness)	Knowledge of idiosyncratic investments; Ongoing knowledge of transaction terms & language; "Bounded knowledge"	Perceived competence of top management; Symbolic status distinctions

it contributes to environmental adaptation and problem solving. Thus, functionalists have established the "themes" of coping with uncertainty, inducing adaptation, and channeling commitment. As functionalist views of culture are often abstract, the discussion of these themes below, will draw on a wider literature for specific content in Table 1.

Coping with Technological Complexity and Uncertainty

Lawrence and Lorsch (1967) found empirical evidence that R&D departments had much less formal structure and much longer time-orientations than other departments. In general, this work clarified two important structural principles: one concerning the "requisite variety," and the other concerning environmental change and necessary levels of integration for adaptation and flexibility.

The principle of requisite variety proposed that an organization needs to be as complex as its environmental suprasystem (Ashby, 1968). This suggests that imbedded innovative units must be culturally unique in order to cope more effectively with its technological slice of its environment. This is evidenced by its highly specialized jargon, values favoring technical sophistication over efficiency or pragmatism, professional allegiances, and so on.

Accelerating environmental change, increasing product complexity and intensified competition demand flexibility. Thus, the imbedded innovative unit must be flexible in dealing with very culturally distinct departments (Miles, 1980). This is accomplished culturally through values and norms that embrace change and establish constructive patterns of interaction.

Autonomous Strategic Behavior

The imbedded innovative unit serves both to reduce uncertainty for strategic planners and to *introduce* it. In describing this relationship, Landau (1985, p. 44) proposed that an R&D unit "exists for the very purpose of upsetting the assumptions made by its strategic planning group." This is functional in that, under conditions of environmental turbulence, the anticipation of the future cannot be based on extrapolation. Consequently, as Ansoff (1985) pointed out, historical strengths may become future weaknesses and ingrained premises can stifle adaptation. Thus, the imbedded innovative unit serves an important strategic role in the strategy formulation process. Burgelman (1985) takes this a step further to suggest that such units engage in "autonomous" strategic behavior. This ability to buck strategy (i.e., as induced by upper management) is bound up in the culture of the unit and is shaped by its interactions with top management and its professional affiliations. Norms that advocate a disregard for organizational procedures (Raelin, 1986) may lead to such autonomous strategic behavior.

Channeling Collective Commitment

From the functionalist perspective, culture provides a vehicle for an implicit understanding and identification with the goals of the social unit, thereby channeling member commitment. This may be through common social values (Durkheim, 1915; Lodge & Vogel, 1987), a shared understanding of effort or outcome contingencies, or through the referent power of its leadership (Jay, 1967). As such, culture builds an emotional identification and pride with the unit's purpose, lessens the propensity to seek employment elsewhere and programs the parameters of behavior and decisions (March & Simon, 1958; Simon, 1987).

Functionalists tend to view cultural cognitions as a type of centripetal force that compensates for egoistic tendencies. Much of the organizational cultural literature, which emphasizes the value of ceremony, symbol and ritual, can trace its genealogy back to Durkheim's (1915) seminal works on a "conscience collective." Given the prevailing and misplaced notions of innovation as an egoistic pursuit, this is an important contribution. More specifically, if an innovative unit has a totemic-like symbol, such as a uniform, building or logo it may heighten collectivism. The building in which the MacIntosh computer was conceived, for example, flew a Jolly Roger pennant.

Table 1 shows examples of cultural content that would be consistent with these functional contributions. For example, heuristic values regarding "keeping current" with the literature would aid in managing uncertainty, and values about the contribution of "bootlegging" (i.e., the quasi-legitimate practice of pursuing independent ideas) may challenge administrative premises. Other cultural cognitions capture the shared knowledge of cause and effect by which unit participants respond. In imbedded innovative units such models may be formalized in scientific theory. However, culture also embodies the shared causal perceptions central to expectancy theories of motivation (e.g., Vroom, 1964). Finally, at the descriptive level, the innovative culture encompasses the knowledge that facilitates communication (e.g., scientific jargon).

Utilitarianism—Transaction Cost Analysis

The primary goal of the utilitarian transaction cost framework (originally proposed by Williamson [1975] and elaborated by Ouchi [1979] and Wilkens & Ouchi [1983]) is to explain the use of alternative forms of governance structures in terms of their comparative efficiency in enabling transactions. Culture, or "clan control," serves as an implicit governance system that is an alternative to market or bureaucratic control mechanisms. From this perspective, cultural control is particularly important for innovative units.

Indeed, it is argued to be the most economically efficient form of transaction governance for frequent and uncertain transactions. They reason that once innovative transactions are brought in-house as a department, market governance is largely forsaken.[2] Of course, the other governance option, bureaucratic control, is not feasible, as it is predicated upon "the assumption that it is feasible to measure, with reasonable precision, the performance that is desired" (Ouchi, 1979, p. 843). This of course, is inconsistent with innovation. Due to the inapplicability of these traditional forms, culture must govern the transactions of innovative units thereby making two important contributions:

Economic Efficiency

From a transaction cost perspective, cultural governance of innovative units is most economically efficient. For example, with frequent R&D transactions the ad hoc transactional inefficiencies of repetitive negotiation would be avoided. For example, an innovative manufacturer of engine seals may need frequent R&D in order to make incremental product improvements. To the extent that R&D in this area becomes ongoing, an internally governed relationship would use fewer resources than external contracting.

Coping with Transaction Uncertainty

Both the uncertainty of transactions and the "bounded rationality" of the parties limit their ability to acquire additional knowledge. Furthermore, the ambiguity of the situation increases the potential for "opportunistic behavior" (i.e., behavior at the expense of the other party). Under these conditions, each party must be somewhat wary of the other's response as new information becomes available over time. For example, with novel research, not only are outcomes harder to anticipate, but the employer faces additional uncertainty about the ability to protect the proprietary knowledge. Thus, the employer must be concerned with the opportunistic behavior of the scientist and would seek available safeguards.

Additional uncertainty is due to the likelihood of idiosyncratic investments (i.e., investments that cannot be transferred to other uses and result in losses if the transaction is unfulfilled). For example, scientists make a personal investment in pursuing the unique interests of the employer. Employers, on the other hand, may purchase expensive instruments at the request of the scientist. This nontransferability creates mutual dependency. The combination idiosyncratic investment and uncertainty prohibit the incorporation of sufficient safeguards in any "contractual" sense and, thus, requires a cultural solution.

According to Wilkens and Ouchi (1983), in such cases, culture reduces uncertainty through the creation of a shared perception that it is to everyone's best interest to cooperate and that, in the long run, everyone will be treated equitably. This requires a foundation of values emphasizing trust, an appreciation for other's contributions, and collaboration. It is observed that, in addition to its obvious functional contributions, much of Organization Development (OD) lays such a foundation for the cultural governance of transactions. While the use of OD in management teams has been widespread and has been used in some innovative contexts (e.g., NASA), its application to R&D units, per se, has received much less attention.

In Table 1 some exemplary values that support utilitarian purposes are proposed. The heuristic that joint effort is the best way to achieve individual self interest (Wilkens & Ouchi, 1983) is central to "clan controls." It is interesting to note the international implications of this value. Hofstede (1982), for example, observed large international differences in "collectivism." It seems reasonable to expect substantial differences internationally in the ease by which such a value may be fostered. Furthermore, Wilkens and Ouchi (1983) also stated that a shared belief that long term justice will be realized is a fundamental basis for cultural governance (i.e., cause/effect knowledge). Finally, a large body of descriptive and symbolic knowledge provides a basis for effective transactions.

Conflict Theory

Conflict theorists have focused on the coalitional relationship of owners and managers to operatives and have used cultural explanations in order to explain the maintenance of certain inequities in social units (e.g., Collins, 1975; Gaventa, 1980). Conflict theory emphasizes that culture is central to the exercise of power, albeit unobtrusively; its principal contribution is in obscuring these inequities (Gaventa, 1980; Lukes, 1974). The potential for widely divergent interests in innovative units is inconsistent with the trust and equity highlighted by the utilitarian perspective. Nevertheless, a potential for divergence exists in the distribution of benefits accruing from an innovation (Pinchot, 1985) and with cleavages in professional vs. managerial values (Raelin, 1986).

Legitimating the Distribution of Benefits

Traditionally, the distribution of benefits has been legitimated in economic organizations in the property rights of its owners. It would seem that this problem would be particularly acute in the imbedded innovative context, because of an ideological tendency to assign ownership rights to the creator of an idea and its applications. However, the current reality is that most imbedded innovative relationships exist between creators of ideas and a group of controlling and nonowning top managers (Berle, 1959) of those ideas. Certainly, in cases where a biased distribution of those benefits deriving from the idea go to managers (and stockholders), some legitimation is required. Contributory components of this cultural legitimation may be found in such shared cognitions as: the use of managerial competence and the "professionalism" of management as a basis for this distribution; that innovative activity, altruism and a "quest for knowledge" is its own reward; and the prominence of the risk/return relationship.

Accommodation of Value Cleavages

As Raelin (1986) described it, there is often a "clash of cultures" between professional employees and managers. This cleavage of values is attributed to the professional's need for autonomy and a concurrent managerial need for control and intervention. Professionals, for example, are trained to look toward their profession for standards and guidance and treat bureaucratic rules as intrusions. Managers, on the other hand, are socialized to articulate goals and structural means for accomplishing them. This suggests that another function of the innovative unit culture is to accommodate those inevitable managerial interventions. For example, innovative cultures generally contain shared cognitions that accept the appropriateness of managerial control over desired "ends," but maintain the belief in control over professional autonomy in establishing the "means" and the importance of informal consultation.

MULTIPLE DIALECTICS

This previous discussion was intended to convey some grasp of the diverse cultural cognitions of an imbedded innovative unit, to clarify the multiple purposes served by its cultural values and to call attention to the contradicting dimensions of these cultures. A closer inspection of Table 1 reveals the presence of at least five major dialectic dimensions where opposing values require synthesis rather than compromise.

Creative Freedom/Control

Creativity, the liberating element of innovation, must be constrained by cultural governance structures. While the absence of control may occasionally lead to serendipity, it more often will result in a technologically driven firm (Ansoff, 1985), grossly overrun expenditures (Marshall & Meckling, 1962), violated timetables (Norris, 1971) and large inventories of unused or unsuccessfully launched ideas. It is clear that, on one hand, the culture of imbedded innovative units must be overtly conducive to creativity, yet must unobtrusively constrain participants on the other.

It is the misdirected efforts of managers to use bureaucratic control mechanisms and a perceived incompatibility of control with "collegial management" that have obscured the importance of control in the innovative process. In Table 1, for example, a value that emphasizes "bootlegging" is a powerful symbol of this freedom; nevertheless, there must be other values that control in the presence of such freedom. Since the notion of control clashes with ideological notions about innovation, presumably cognitions of control will be more unobtrusive and involve broad parameters of decision and action. In this respect, it has been observed that mission statements serve as much to clarify what not to pursue rather than by providing specific direction (Ansoff, 1985). For example, the concept of two-dimensionality guided new product development at 3M Corporation (Peters & Waterman, 1982), thereby creating well-defined parameters for creative activity. The synthesis of freedom and control appears to lie in culture's ability to restrain egoism, and to establish unobtrusive control parameters within which a more overt freedom values emphasized, and a team orientation in its membership founded on trust and equity.

Differentiation/Integration

The distinctiveness of the R&D department is a recurrent theme. Nevertheless, the successful management of technology requires integration with other diverse units. Ansoff (1985, p. 120), for example, used the term "downstream coupling" to refer to the "blending and integration of functional contributions" with technological development. Although earlier theorists emphasized marketing (e.g., Schmookler, 1966; Freeman, 1982), there is a pressing need for effective integration with other functions, such as production and design. The distinctiveness of the innovative unit needs to be accompanied by an acquaintance with the other function's cultures. As Miles (1980, p. 261) pointed out, the difficulty is obtaining integration without homogenization: "Bureaucratic attempts to integrate units dealing with changing environments tamper with their requisite differences." This synthesis is possible through the use of an integrating "corporate culture" and other "nonconventional" integration mechanisms (Lawrence & Lorsch, 1967) that emphasize building communication and further cultural bonds through intraorganizational mobility.

Individualism/Collectivism

The imbedded innovative unit should be characterized by "creative individualism" (Jay, 1967) and managerial leadership (Schein, 1985).

Relatedly, Durkheim's (1933) discussion of "organic solidarity" seems to be trying to make a similar point. The issue in this treatise was how a markedly differentiated society can confront "anomie" (i.e., normality). In part, his answer involved values imparting a "cult of the individual." The Durkheimian notion of individualism is complex and it is merely noted here that it involves strengthening bonds through self-actualization rather than egoism. In innovative units, this synthesis may also be promoted through norms and reinforcing reward systems associated with professional activity and publication. On the other hand, the need for a cross fertilization of ideas and interdependencies required for commercial development require cooperation.

Localism/Professionalism

Gouldner (1957) first identified a tension between professional and corporate loyalties. Professionals are inculcated with values that are consistent with the aims of the overall profession as well as its political advancement (Clark, 1939; Raelin, 1986). Yet, as a part of a specific firm, it is important that this same scientist be locally oriented. Of course, a synthesis is facilitated in that, for the most part, professional norms are consistent with good practice. Thus, innovative managers need to work with scientific or artistic subordinates in order to guide and refine these norms and values to fit the local context. One part of this tension involves the previously discussed professional need for autonomy and a managerial need for control.

Contemplation/Action

The creative process is characterized by reflection and incubation (Kolbe, 1975)—a process that would seem to be recalcitrant to acceleration. However, under current competitive conditions the need to market new technologies is more and more pressing. Jay (1967, p. 107) points out, however, that often "the knowledge that a deadline is approaching, that something has to be done urgently, is a wonderful liberator of the creative impulse." Jay (1967) also relates this to the R&D manager, stating that he or she must be both a reflective and inspirational "yogi," as well as an administrative "commissar." This imagery suggests the presence of other dialectics in innovative units: idealism versus realism; introspection versus judgment; wholeheartedness versus tentativeness; and sophistication versus pragmatism.

This section demonstrated that the culture of imbedded innovative units is inherently dialectical. In addition to technical credibility, the effective management of these units appears to lie in synthesizing, rather than compromising, these multiple contradictory cultural cognitions. Effective R&D management involves a good deal more than splitting the differences. In other words, the most effective R&D manager is one who can attain both freedom and control, localism and professionalism, and so forth.

IMPLICATIONS

The modern manager of an imbedded innovative unit faces a formidable task—the maintenance of technological sophistication, the attainment of high levels of effective integration with other functions and the management of a cultural synthesis of these multiple contradictions. Barney (1986) has argued that a "corporate culture" can only realize a competitive advantage to the extent that it is valuable, unique and imperfectly imitable. A full theoretical development of which contingencies will have the greatest impact on the culture of the innovative unit is beyond the purpose of this paper. Nevertheless, from the previous discussion of perspectives it would appear that the need to achieve a cultural synthesis of these theoretical dimensions will be contingent upon the ability to operationalize the firm's strategy. Of course, this relates to how central the innovative unit is to the performance of the firm and how innovative the unit must be.

Proposition 1. The importance of achieving a synthesis among these multiple dialectical dimensions is associated with the capability of operationalizing a firm's strategy.

The arguments in this paper suggest that the attributes of "synthesized" and "integrated" be added to this list. Furthermore, it seems evident that if these qualities are realized, value and uniqueness must follow.

Proposition 2. The realization of a competitive advantage from innovation in larger organizations will be related to the level of synthesis of multiple dialectical dimensions achieved through the culture of its imbedded innovative units.

Contingencies

A cultural synthesis will be dependent upon many contingencies associated with each unit's situation. For example, the manager of an imbedded innovative unit must be sensitive to a host of influences, such as: the centrality of innovation to the strategy of the business(es); the unit's political standing; the professional ambitions of the staff; the values of upper management, and so on. Consequently, any universal prescriptions, such as those mentioned at the beginning of this paper, must be viewed cautiously and with the realization that for each prescription there will be a companion, dialectical value. A full theoretical development of which contingencies will have the greatest impact on the culture of the innovative unit is beyond the purpose of this paper. Nevertheless, from the previous discussion of perspectives it would appear that the need to achieve a cultural synthesis of these theoretical dimensions will be contingent upon the ability to operationalize the firm's strategy. Of course, this relates to how central the innovative unit is to the performance of the firm and how innovative the unit must be.

Critical Introspection

A change in the management of these dialectical cultural cognitions needs to begin at home. Management must understand its own culture, biases, contradictions and egoistic tendencies. Lodge and Vogel (1987) argue that the prevailing ideology within the United States may be ill-suited to the realities of international competition. They cite, among others, the American preoccupations with egoism as a basis for fulfillment and specialization as ideological impediments to adaptation to new competitive realities of the global marketplace. This is particularly troublesome as a mismatch of ideology from its economic context is insidious, difficult to diagnose and has a poor prognosis. Furthermore, an accurate diagnosis is easily rejected by the patient and projected onto other causes.

At the firm level, the most pragmatic approach to this difficulty probably lies in careful selection and location. Managers and members of imbedded innovative units need to be appraised of their value profile and their ability to embrace contradictory values. This may be stated as the following proposition.

Proposition 3. Members and especially managers of effective imbedded innovative units will be more comfortable in embracing and synthesizing contradictory values than members of less effective units.

Cognitive Development

With considerable effort it may be possible to impart cognitive frameworks that may be more compatible with synthesis. Perry (1970), for example, described a model of cognitive and ethical development whereby individuals are believed to adopt successively more sophisticated and complex frameworks for interpreting reality. According to Perry, individuals come to such a position through a progressive intellectual journey characterized by stable stages separated by periods of transition. Prior to reaching the highest level of cognitive sophistication, individuals are unable to commit to actions by incorporating internally derived ethical values with a full appreciation of a dialectical situation.

Proposition 4. Controlling for technical competence, managers of more effective imbedded innovative units will be more cognitively sophisticated and ethical than managers of less effective units.

Building Trust

It is likely that managers may be trained, for example, to broaden their conceptualizations of control in order to appreciate that the management of such units does not amount to abdication or to bureaucratic intrusion. As previously mentioned, from a utilitarian perspective the foundations for cultural control are a shared belief that joint effort is the best way to achieve individual self interest, equity and long term justice (Wilkens & Ouchi, 1983). Of course, this cannot be created by innovative managers alone, but must cascade down from the values, action and leadership of top management. This leads to the following three propositions.

Proposition 5. There will be a greater emphasis placed on broader cultural criteria in the selection process in more effective innovative units.

Proposition 6. There will be higher levels of congruity regarding values of trust within the unit and with top management in effective innovative units than in less effective units.

Proposition 7. There will be a higher levels of perceived procedural and distributive justice within effective R&D units than in less effective units.

Synthesizing Managerial and Technical Education

One major difficulty in embracing dialectical reality in the traditional R&D setting is its incomprehensibility within the overall scientific or engineering paradigm. A rigid technical curriculum based on elegant transformations to correct solutions is ineffective preparation for developing useful frameworks for social realities.

While some success may be accomplished through in-house training and prudent selection, the rejuvenation of corporate innovative processes needs to be grounded upon a broader educational foundation that goes beyond imparting factual knowledge of physical properties. Scientists and managers alike need to develop an intellectual capacity for appreciating contingencies, introspection, an appreciation of the existence of dialectical truth, as well as ethical sophistication. As long as such training is needed at the corporate level to compensate for institutional ineffectiveness, domestic corporations will be at a competitive disadvantage. This suggests a need for institutions to broaden their curriculum. Of course, this is facilitated to the extent that education is supported by other social institutions. Obviously, extremely important institutions for shaping a nation's values are its religions. The intransigence of religions probably must be taken as a given; however, it is observed that religions such as Taoism (with its "yin and yang" conceptualization of reality) are compatible with dialectical cognitive frameworks.

CONCLUSIONS

This paper explored the culture of imbedded innovative units and found the universalistic values and norms that have been proposed in the literature to be inadequate. It was found that these prescriptions were often inconsistent or seemed to be overly simplistic. An inspection of the content of innovative cultures permitted the identification of the inherent dialectical nature of innovative management in larger firms. Difficulties in managing R&D were associated with transcending a simple compromising approach to these values toward a synthesis in a dialectical sense. Freedom is illusory in anarchy; creativity bears little fruit without governance; and, control atrophies in the absence of change. One particularly interesting point that emerged from these arguments is a need for the creation of a culture of equity and justice in the management of innovative units. In imbedded innovative units, this provides a necessary, but not sufficient, basis for multiple functions: the governance of the employment relationship, integration, autonomous strategic behavior, and even for "efficiency." This rationale for ethical behavior needs to be developed and the relationship of justice in innovative units to its performance is an exciting area for further research.

It is suggested that the prescription for our domestic industrial R&D must include a systemic and broad-spectrum potion. In the case of innovation, such cultural changes must involve institutions that inculcate these values to professional scientists and engineers. Furthermore, mere topical applications will not penetrate to the depths of outmoded managerial ideologies. These multiple and apparently contradictory dualities clarify that high levels of sensitivity and cognitive development are required for the effective management of innovative units.

NOTES

1. Anthropological and phenomenological conceptualizations of organization culture are acknowledged to be useful in understanding the evolution and enactment of culture by participants; however, they are less clearly linked to explicit organization level goals. Consequently, the contributions of an imbedded innovative unit are difficult to determine.

2. From a utilitarian perspective, many of the recommendations in the recent literature on "intrapreneurship" (Pinchot, 1985) can be viewed as an attempt to reintroduce market controls on these transactions. While some of these measures may be useful, they are inadequate in the imbedded context considered in this paper and require some level of cultural governance.

REFERENCES

Ansoff, H.I. (1985). *Implanting strategic management.* Englewood Cliffs, NJ: Prentice-Hall.

Ashby, W.R. (1968). Variety, constraint, and the law of requisite variety. In W.B. Buckley (Ed.), *Modern systems research for the behavior scientist* (pp. 129-136). Chicago: Aldine.

Barney, J.B. (1986). Organization culture: Can it be a source of competitive advantage? *Academy of Management Review, 11*(3), 656-665.

Berle, A.A. (1959). *Power without property: A new development in American political economy.* New York: Harcourt, Brace & World.

Burgelman, R. (1985). Corporate entreprenuership and strategic management: Insights from a process study. *Management Science, 29*(12), 1349-1364.

Burns, T., & Stalker, G.M. (1961). *The management of innovation.* London: Tavistock.

Clark, J. (1939). *Social control of business.* New York: McGraw-Hill.

Collins, R. (1975) *Conflict sociology: Toward an explanatory science.* New York: Academic Press.

Competitiveness Survey: HBR readers response. (1987). *Harvard Business Review, 65*(6), 8-12.

Dean, R. (1974). The temporal mismatch: Innovation's pace vs. management's time horizon. *Research Management, 17*(3), 12-15

Durkheim, E. (1915). *The elementary forms of religious life* (J.W. Swain, Trans.). New York: Free Press.

________(1933). *The division of labor in society* (G. Simpson, Trans.). New York: The Free Press.

Dyer, W. (1985). The cycle of cultural evolution in organizations. In R.H. Kilman, M.J. Saxton, & R. Serpa (Eds.), *Gaining control of the corporate culture* (pp. 200-229). San Francisco, CA: Jossey-Bass.

Ellul, J. (1964). *The technological society.* New York: Summit Books.

Foster, R. (1986). *Innovation: The attacker's advantage.* New York: Summit.

Freeman, C. (1982). *The economics of industrial innovation* (2nd ed.) Cambridge, MA: M.I.T. Press.

Gaventa, J. (1980). *Power and powerlessness: Quiescence and rebellion in an Appalachian Valley.* Urbana, IL: University of Illinois Press.

Gold, B. (1967). *The decision framework in generating major technological innovations: Some unfashionable hypotheses* (Working paper no. 6). Cleveland, OH: Case Western Reserve University.

Hayes, R.H. & Abernathy, W.J. (1980). Managing our way to economic decline. *Harvard Business Review, 58*,67-77

Hofstede, G. (1982). *Culture's consequences.* Beverly Hills, CA: Sage.

Jay, A. (1967). *Management and Machiavelli: An inquiry into the politics of corporate life.* New York: Bantam.

Kay, N.M. (1979). *The innovative firm: A behavioral theory of corporate R&D.* New York: St Martin's Press.

Kolbe, D.A. (1975). Four styles of managerial learning. In D.T. Hall, D.D. Bowen, R.J. Lewicki, & F.S. Hall (Eds.), *Experiences in management and organization behavior* (pp. 19-50). Washington, DC: American Chemical Society.

Landau, R. (1980). Chemical industry research and innovation. In. W.N. Smith & C.F. Larson (Eds.), *Innovation and U.S. research: Problems and recommendations* (pp. 19-50). Washington, DC: American Chemical Society.

Lawrence, P.R., & Lorsch, J.W. (1967). *Organization and environment: Managing differentiation and integration.* Homewood, IL: Irwin.

Lodge, G.C., & Vogel, E.F. (1987). *Ideology and national competitiveness: An analysis of nine countries.* Boston: Harvard Business School Press.

Lucas, R. (1987). Political-cultural analysis of organizations. *Academy of Management Review, 12*(1). 144-156.

Lukes, S. (1974). *Power: a radical view.* London: Macmillan.

March, J.G., & Simon, H.A. (1958). *Organizations.* San Franciso CA: Wiley.

Marshall, A.W., & Meckling, W.H. (1962). *Predictability of the cost, time, and success of development.* Washington, DC: National Bureau of Economic Research.

Miles, R.A. (1980). *Macro organizational behavior.* Santa Monica CA: Goodyear.

Miller, M.W. (1987, September 19). High tech world sees IBM case as a way out of the copyright maze. *Wall Street Journal,* p. 1.

Norris, K.P. (1971). The accuracy of project cost and duration estimates in industrial R&D. *R&D Management, 2*(1), 25-36.

Ouchi, W.G. (1979). A conceptual framework for the design of organizational control mechanism. *Management Science, 25*(9), 833-848.

Perry, W.G. (1970). *Forms of intellectual and ethical development in the college years.* New York: Holt, Rinehart, & Winston.

Peters, T., & Waterman, R.H. (1982). *In search of excellence.* New York: Harper & Row.

Pinchot, G. (1985). *Intrapreneuring: Why you don't have to leave the corporation to become an entrepreneur.* New York: Harper & Row.

Raelin, J.A. (1986). *A clash of cultures: managers and professionals.* Boston: Harvard Business School Press.

Rawls, J. (1971) *A theory of justice.* Cambridge MA: Harvard University Press.

Schein, E.H. (1985) *Organizational control and leadership.* San Francisco CA: Jossey-Bass.

Schmookler, J. (1966). *Invention and economic growth.* Cambridge CA: Harvard University Press.

Skinner, B.F. (1971). *Beyond freedom and dignity.* New York: A.A. Knopf.

Simon, H.A. (1987). Making management decisions: The role of intuition and emotion. *Academy of Management Executive, 1*(1), 57-64.

Smircich, L. (1983). Concepts of culture and organizational analysis. *Administrative Science Quarterly, 28*(3), 339-358.

Vroom, V.H. (1964). *Work and motivation,* New York: Wiley.

Werner, O., & Schoepfle, G. (1979). *Handbook of ethnoscience.* Evanston, IL: Department of Anthropolgy, Northwestern University.

Williamson, O. (1975). *Markets and heirarchies: Analysis and antitrust implications.* New York: Free Press.

Wilkens, A., & Ouchi, W.G. (1983). Efficient cultures: Exploring the relationship between culture and organizational control. *Administrative Science Quarterly, 28,* 468-481.

LOCAL KNOWLEDGE SYSTEMS IN ADVANCED TECHNOLOGY ORGANIZATIONS

Marietta L. Baba

Technology and structure are two fundamental dimensions of formal organization and an understanding of their relationship is crucial to the successful management of advanced technology firms. Research on the technology-structure interface has been concerned primarily with the possibility of causal relationships between the formal aspects of technology and structure; that is, the consciously planned and deliberately executed input-output flows that constitute formal technology and the arrangements of people and subunits that comprise formal structure within an organization (Fry, 1982). Less frequently, such research has considered the informal aspects of technology and structure; that is, the relationship between unplanned and spontaneous organizational processes and human interactions (see for examples Burawoy, 1979; Vaverek, 1987).

Technology has both a formal and an informal component, and the informal component (referred to herein as *local knowledge*) is a by-product of interaction between formal technology and informal organization (see Figure 1). Local knowledge is created by informal work groups in response to technological capabilities and limitations that were not envisioned by the designers of technology, but which emerge and command attention under conditions of actual use. To the extent that informal knowledge systems reflect both the inherent potential and the limitations of formal technology, it is argued here that local knowledge is shaped in important ways by a technological imperative—an imperative mediated by the creative problem-solving actions of informal social organization.

This chapter extends earlier work on the content and structure of informal knowledge systems in traditional manufacturing organizations by suggesting that local knowledge takes on a special character at advanced technology work sites (defined here as bounded work domains associated with advanced technology processes and/or products that are key to an organization's competitive position). The special characteristics of local knowledge systems at such sites include a distributed (i.e., widespread network) structure, creative expansion over time, and a content that includes information pertaining to systemic interrelationships. These characteristics both accommodate and reflect the special character of advanced technological processes and/or products. The chapter argues that local knowledge systems are an integral part of organizational technology and key contributors to productivity in advanced technology organizations. Managers in such organizations are advised to map the location of local knowledge systems and preserve their integrity in planning for organizational and/or technological change.

Many of the studies cited in this chapter are drawn from the tradition of ethnographic research; that is, empirical inquiry whose aim is understanding and interpreting behavior across systems of meaning (i.e., cultures). Most frequently, such ethnographic research has involved direct observation of behavior in natural field settings. Because local knowledge is informal, and oftentimes illicit, it has proven resistant to more structured data acquisition strategies and thus better suited to qualitative research in the ethnographic tradition. The relative paucity of empirical studies focusing on local knowledge systems in advanced technology organizations, however, means that the propositions set forth in this chapter must be considered tentative and preliminary.

DEFINING AND CHARACTERIZING LOCAL KNOWLEDGE

For purposes of this chapter, *local knowledge* is defined as a complex system of shared information, including abstract models of reality and methods of problem-solving related to technology, which is not formalized but is created spontaneously among work group members, and is used by group members to support the performance of work tasks.[1] The word *local* in this context is used to indicate that the knowledge is localized (i.e., contained) within an organizational subunit or system of subunits whose boundaries can be specified. Local knowledge is learned and shared by group members, and is transmitted to new members through informal communication networks whose structure can be mapped by social network analysis.

The Neoclassical Tradition

The existence of local knowledge systems and their effects on production first were hinted at indirectly in classic observational studies of workforce behavior in traditional manufacturing settings. In the Hawthorne experiments, Mayo's colleagues noted that some workers in the Bank Wiring Observation Room were able to achieve extremely high work speeds—speeds that seemed to defy what was possible by formal engineering standards (Roethlisberger & Dickson, 1939) Later, Roy (1952, 1954, 1958) showed that high speeds derived in part from illicit procedures used by machine operators to "make out" on piecework (i.e., produce a large number of pieces on some jobs, while restricting output on others). Working as a radial drill operator in the machine shop of a steel processing plant, Roy discovered that workers knew informal procedures and techniques that could be used to embellish or streamline machine operations before and after time-study ratings. Roy focused his attention primarily on illicit manipulation of company regulations as a means of increasing production speed. He found, for example, that workers sequestered main set-ups (i.e., basic tools needed to perform a range of jobs) under their work benches (rather than turning them in according to company regulations) in order to save set-up time on new jobs (Roy, 1954). More recent studies in the neoclassical tradition show that factory workers also perform illicit manipulations on machines themselves in order to boost work speeds. For example, Shapiro-Perl (1979) found that workers in a costume jewelry shop streamlined production by piling and processing together certain jewelry components in order to attain high volume production. Lamphere (1979) similarly discovered that individual workers in the apparel industry invented special sewing tricks that enabled them to increase production (and which they also taught to work trainees).

Unfortunately, none of the observational studies cited above nor others conducted during the 1940s and 1950s (e.g., Gouldner, 1954; Homans, 1953; Whyte, 1948) provide a detailed description or analysis of the informal knowledge base that mediates human interaction with the labor process. The focus of industrial sociology and anthropology traditionally has been the informal social relations of production—not the informal knowledge base that both underlies, and results from, social and technical relations at the work place. As a result of researchers' overriding interest in social relations, the informal aspects of production generally have been couched in social terms. Most typically, informal work procedures have been viewed as an extension of the informal work group's social control over production, with their function tied specifically to the promotion of internal group cohesion (see Blau & Scott, 1962). The effects of such procedures on production have been characterized generally as running counter to management's productivity objectives (see for discussion Albrecht & Goldman, 1985).

Kusterer's Characterization

One of the first (and most detailed) explicit conceptualizations of informal working knowledge (referred to herein as local knowledge) was given by Kusterer (1978), who conducted case studies of informal technical know-how at various work sites. Kusterer's data collection procedures includes both structured and semi-structured interviews, as well as direct field observation. From his early experience as a printer, Kusterer became convinced that the cognitive apparatus of skilled craftspeople includes knowledge paradigms (in Kuhn's [1970] sense) that organize their perception of work tasks and behavior on the job. Further, Kusterer believed that in the practice of work, skilled craftspeople function much as scientists—solving problems, acquiring knowledge and pushing at the frontiers of their paradigms—all in an informal fashion. Kusterer wanted to determine whether knowledge paradigms exist for all forms of work, regardless of skill level. Thus, he chose for his study occupations requiring relatively low levels of skill.

In Kusterer's (1978) study of the paper cone fabricating department in a traditional manufacturing firm, the labor process was highly mechanized and jobs had been divided into the lowest skill levels possible. Of the five job categories established to handle production, machine operators were the most deskilled (in Braverman's [1974] sense). Thus, Kusterer chose that job group for the focus of his work.

Apart from the formal knowledge acquired by machine operators during their brief training periods (i.e., knowledge related to basic work procedures, safety and quality standards, and company rules), Kusterer identified four types of supplementary knowledge—informal techniques and procedures needed to solve problems related to work performance. Each of these four types are named and described briefly below:[2]

1. *Knowledge about Materials.* Information pertaining to material defects and their impact on machine operations, and informal procedures designed to compensate for such defects (e.g., hand-finishing of materials in an otherwise automated process, using illegal [and unsafe] cleaning methods while machines were in operation).
2. *Knowledge about Machines.* Information pertaining to general machine failures and the idiosyncrasies of particular machines, and informal procedures designed to prevent failure or affect adjustments

that improve productivity (e.g., using machine operators' own tools to make nonstandard alterations of machinery).[3]

3. *Knowledge about Quality Standards.* Information pertaining to several different (and partially contradictory) sets of quality standards (including standards used by inspectors, managers, customers and the work group itself), and the appropriate use of such standards to achieve various production objectives.
4. *Knowledge about the Work Community.* Information pertaining to the social values and norms held by members of other job categories, and the appropriate adoption of such values and norms to enlist aid from members of these other groups (e.g., nurturing an image of competence to encourage aid from mechanics).

The content of these four categories suggests that local knowledge systems are highly focused and selective, relating directly to workers' experiences with technology and containing both technical and social concepts needed to complete specific jobs. Based on his data, Kusterer (1978) suggests that local knowledge systems are most likely to develop around work phenomena that display a capacity for variance (i.e., error or discrepancy), occur with reasonable frequency (i.e., are neither constant, nor rare), and affect the performance of individual work tasks. Given these conditions, it seems likely that opportunities for the creation of local knowledge will vary directly with the number of productive functions assigned to each job (since a larger number of functions should display a larger number of variances). Further, such opportunities should vary with the degree of routinization that has been designed into each job function (since a high degree of routinization is typically associated with a lower degree of variance). In general, therefore, bodies of local knowledge should be more extensive where job functions are more complex and less routinized. It is possible, however, to imagine fairly extensive bodies of local knowledge developing in situations where simpler, more routinized jobs are associated with antiquated machinery and/or defective materials (as in Kusterer's cone fabricating department).

LOCAL KNOWLEDGE: A TECHNOLOGICAL IMPERATIVE?

Kusterer's (1978) pioneering work both confirms and extends the portrait of informal work procedures drawn by neoclassical scholars. Confirmation is provided for the neoclassical view of informal procedures as spontaneous, cooperative activities that enhance work group cohesion (and, in Kusterer's view, reduce alienation). Additionally, evidence is marshalled to support the premise that informal work procedures are illicit or underground phenomena (i.e., they are hidden from management). In the cone fabricating department and among other occupational groups studied by Kusterer, the creation and use of local knowledge systems often required work groups to violate formal job descriptions, make unscheduled alterations of machinery, engage in unsafe maintenance practices and ignore an assortment of corporate rules and regulations concerning material handling.

Even more importantly, however, Kusterer's work shows that informal work groups both affect technology through social means of control, *and* interact creatively with technology—discovering its hidden potential and coping with its emergent limitations in ways that generate informal bodies of knowledge related to technology. Indeed, from Kusterer's research we see that bodies of informal knowledge develop directly in response to gaps or inadequacies in formal bodies of knowledge pertaining to technology. This argument is elaborated as follows.

We know that designers of formal technology (whether material or social) typically cannot predict exactly how a given piece of technology will perform on the shop floor or office environment (Doutt, 1959; Tornatzky, Eveland, Boylan, Hetzner, Johnson, Roitman, & Schneider, 1982; Tornatzky & Fleischer, 1990). The designers may not know, for example, all of the ways that the technology can fail (e.g., when the technology is antiquated or is interacting with defective material). They also may not know all of the techniques that can be used to prevent or correct failure, or to otherwise maximize production under high pressure conditions. As a result of these gaps in formal bodies of knowledge, the formal training or documentation which accompanies technology often is inadequate to the task of operating such technology under actual conditions of use (Orr, 1986a). Work groups whose members confront technology directly under real operating conditions, on the other hand, are in a position to gain experience regarding its performance and it is these work groups whose members must cope with emergent technological problems in order to do their jobs. Work groups learn the real capabilities and limitations of technology through use, and in response to this reality, they informally create new ways to maximize the technology and/or to cope with its inherent limitations (see for examples Howard & Schneider, 1988; Orr, 1986a; Shimada & MacDuffie, 1986).

The present conceptualization suggests that local knowledge, although at times illicit, actually may support or enhance production and is in fact an integral part of organizational technology. Just as informal organization is an important component of any organizational system and arises "in response to the opportunities created and problems posed by the environment" (Blau & Scott, 1962, p. 6), so local knowledge may be viewed as a crucial component of any technological system—one that also arises in response to opportunities and problems posed by the technological

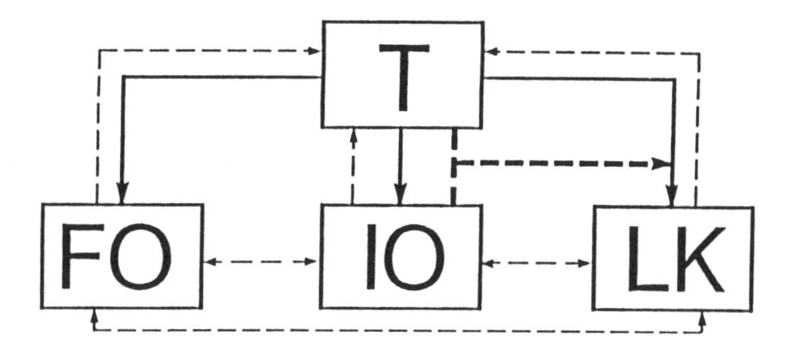

Source: Adapted from Burawoy (1979), Fry (1982), Kusterer (1978), Vavarek (1988), and Woodward (1965).

Figure 1. Technology—Stucture Relationship
An Interactive Conceptual Model

environment. It seems reasonable, then, to argue that the content and structure of local knowledge systems may be shaped by a sort of technological imperative—a set of implicit technological capabilities and limitations that can only be discovered and overcome as technology operates under actual conditions of use. While such discoveries and solutions are undoubtedly mediated by informal social mechanisms (e.g., troubleshooting sessions among informal work groups; see Sachs, 1988), their fundamental content is a mirror of inherent technological properties. The foregoing argument may be expressed as a formal proposition, stated as follows:

Primary Proposition 1. Technological systems embed an informal component which results from, and reflects, the inherent potential and the limitations of formal technology.

The proposed theoretical position of local knowledge relative to formal technology (and to other dimensions of organizational structure) is depicted graphically in Figure 1. The solid lines and unidirectional arrows in this diagram portray primary determinative (or causal) relationships as they have been characterized in the literature of modern organizational theory.

Technology (in the formal sense) is represented as a primary determinative force in the structuring of both formal (Fry, 1982) and informal organization (Burawoy, 1979; Vavarek, 1987). The diagram also suggests that technology—its potential and its limitations—is a primary motive force driving the creation and content of local knowledge systems.

Mutually supportive and interactive relationships among the technological and structural dimensions of organization are represented in the diagram by broken lines and bi-directional arrows. These interactive relationships include the known influences of formal organization on formal technology and informal organization, documented feedback effects from informal organization to formal structure (Jewell & Reitz, 1981), documented interactions between informal organization and local knowledge (Orr, 1986a), and the probable (but as yet undocumented) relationship between formal organization and local knowledge. The special interaction of formal technology and informal organization which creates bodies of local knowledge (Kusterer, 1978) is represented by a heavy broken line. The need for further research to explore some of the various interrelationships suggested by Figure 1 is considered in the discussion section of this chapter.

LOCAL KNOWLEDGE SYSTEMS IN ADVANCED TECHNOLOGY ORGANIZATIONS

The suggestion of a technological imperative shaping the content of local knowledge leads to the possibility that informal bodies of knowledge may take on a special character in advanced technology organizations. Such organizations are, by definition, dependent on advanced (i.e., high) technology processes and/or products for their competitive positions (see for discussion Balkin & Gomez-Mejia, 1984; examples of advanced technologies discussed later in this chapter include automated inventory planning and control systems, and electronics-based components and end-user products). Advanced technologies, in turn, are complex and rapidly changing phenomena that embody relatively large and expanding domains of formal knowledge. It follows, then, that advanced technology organizations should encompass informal bodies of knowledge that are proportionately more complex and rapidly evolving than those located in traditional organizations. Complex and rapidly evolving systems of local knowledge will be contained specifically (i.e., localized) within advanced technology work sites; that is, bounded work domains associated with advanced technology processes and/or products that are key to an organization's competitive position. This argument may be summarized in a second formal proposition, set forth as follows:

Primary Proposition 2. Local knowledge systems will be relatively more complex and rapidly evolving at advanced technology work sites than at traditional work sites.

This proposition derives from our current understanding of local knowledge as a body of information that responds to and reflects the potential and the limitations of formal technology. Where formal technology is relatively more complex and rapidly changing, the structure and content of local knowledge systems should reflect this complexity and evolutionary change.

It is the rate of technological change that is especially important for understanding local knowledge systems in advanced technology firms. In order to achieve and maintain a competitive position, the key processes and/or products of such firms must coevolve with their technological bases. New developments in process and product technologies within the firms' external (or internal) environment generate corollary technological change within the firm itself—change that brings new problems and limitations (i.e., variances) that must be solved under actual conditions of use. The newness of advanced technologies means that their variances are potentially infinite in nature; that is, the exact character and frequency of variances cannot be predicted in advance. If local knowledge is shaped in some measure by the requirements of formal technology, then we would expect a potentially infinite universe of variances to be reflected by an ever expanding pool of informal coping responses.

The situation described above contrasts sharply with the nature of problems encountered (and solutions enacted) in traditional organizations. In such organizations, technological bases are mature and key process/product technologies are relatively stable over time. In Kusterer's (1978) cone fabricating department, for example, the products being produced were relatively simple (i.e., paper cones for food products) and the production machinery was somewhat antiquated (i.e., 20-years-old). In such traditional technological environments, the number of variances is finite; members of the work group know the various types and frequencies of technological problems that are most likely to occur in the work place. Further, solutions to the vast majority of these problems are known to the majority of the workforce (with the exception of new recruits). The initial establishment of a traditional manufacturing organization may, of course, involve a fair number of unpredicted variances as part of the start-up process. Eventually, however, start-up problems are resolved and the typical variances encountered are those that are known in the industry.

While an individual machine operator might hope to master eventually all (or virtually all) of the regular machine/material variances encountered and solutions enacted at a traditional work site, workers in advanced technology organizations have no such hope. For these latter workers, processes and products are moving targets, constantly changing to keep pace with technological developments. No individual worker can hope to master—via direct experience—all of the ways there are for technology to fail or all of the possible means to cope with technological breakdowns. It is only the total community of workers who together, as a group, experience the total spectrum of technological shortcomings which manifest themselves in the complex and rapidly evolving technological environment. And it is the total community of workers—although perhaps dispersed geographically and/or across long-linked processes—who must devise some means to capture new data gained by individuals and to pool these new data for collective use.

From the foregoing discussion, it is possible to derive three subsidiary propositions (i.e., subsidiary to Primary Proposition 2) concerning the nature of local knowledge systems in advanced technology organizations. Each of these subsidiary propositions (as well as the primary propositions) are supported by empirical evidence drawn from recent ethnographic studies of local knowledge systems at advanced technology work sites (Howard & Schneider, 1988; Orr, 1986a, 1986b, 1988; Sachs 1988). The subsidiary propositions, together with a brief discussion and summary of supporting evidence, are set forth below.

Subsidiary Proposition 1. Individuals and/or informal work groups at advanced technology work sites will create new knowledge in response to technological change, and will contribute such knowledge to an expanding knowledge pool via social mechanisms that enable pooling and sharing of information.

Evidence for the creation and pooling of new informal knowledge has been documented in recent ethnographic studies of technological change. Howard and Schneider (1988) and Sachs (1988) investigated the implementation of automated inventory planning and control systems (Manufacturing Resource Planning—MRP) at an aircraft instruments plant and an electronic components plant, respectively. In both studies, researchers found that MRP (and MRP II) had been designed initially on the basis of an idealized formal representation of reality that did not incorporate informal technical knowledge related to work place production. As a result of this separation of formal and informal knowledge, the automated systems did not work properly in the real-world environment of the plants under study. At the aircraft instrument plant, for example, MRP II would not permit parts to move backward in the production chain, even when components needed to go back to earlier stages for rework (Howard & Schneider, 1988). To cope with such problems, workers developed a new body of informal knowledge that enabled them to understand MRP requirements and deficiencies, and to manipulate around its rules (thereby reducing costly production bottlenecks and delays). New knowledge about MRP was created and shared during informal shop floor conferences between production workers and inventory control personnel.

A similar process of creating and sharing an expanding pool of knowledge was found by Orr (1986a, 1986b) in his study of Xerox service technicians. The newness and complexity of advanced photocopier models prohibited the development of formal training and documentation that anticipated all of the serious problems that might be encountered by service technicians in the field. New models simply had not been in operation by end-users long enough for designers to know the full range of problems that would emerge over time. Thus, technicians frequently encountered problems that had not been predicted previously or addressed in formal training. Orr (1986a, 1986b) found that when routine repair procedures failed to affect satisfactory results on advanced machines, technicians begin to tell stories about past machine failures. These anecdotes combined information about the machine with the context of a specific situation, allowing technicians to compare the symptoms and diagnostic test results described in narration with new situations encountered in the field. Technicians avidly exchanged narratives (war stories) at every available opportunity (e.g., during breaks at the training sessions), thereby creating a communal memory of problems and solutions. According to Orr, Xerox service representatives share abstract mental models of machines that incorporate formal information gained in training and informal information shared through narratives. When a technician faces an intractable problem in the field, the mental models (partially derived from narratives) suggest tests that should be performed when certain symptoms are encountered. Mental models and communal narratives aid in the organization of new knowledge and serve as a framework for the retrieval of information.

A second subsidiary proposition that may be derived from our general theoretical orientation concerns the structure of local knowledge systems at advanced technology work sites. Due to rapid technological change, local knowledge systems at such sites may be expected to display a distributed (i.e., dispersed or widespread) character over time and space. That is, the content of local knowledge will not be possessed in its entirety by any single individual, but will be dispersed across a work group with each individual possessing varying amounts of information. This possibility exists because rapid change will not permit an individual worker to experience directly all (or the majority) of new technological problems or failures that arise over time. Rather, individual workers or informal work groups will experience and cope with new problems in a decentralized fashion, as problems arise unpredictably over time and space. If local knowledge is distributed at advanced technology work sites, then we also might expect relatively greater variability in the amount of local knowledge possessed by any individual worker. While it is true that individual workers in traditional organizations also possess varying degrees of informal knowledge, it is argued here that such variability in advanced technology organizations is more a function of technological change than of individual aptitude or interest. As a result of this difference, the total system of local knowledge at advanced technology work sites probably cannot be obtained from any single worker (no matter how experienced or talented). This argument is summarized in the following subsidiary proposition, which also derives from Primary Proposition 2:

Subsidiary Proposition 2. Local knowledge systems at advanced technology work sites will display a distributed (i.e., widespread network) structure, with total knowledge content dispersed unevenly among individual workers.

Evidence for the distributed character of local knowledge systems at advanced technology work sites is suggested in results presented by Howard and Schneider (1988), and Orr (1986a, 1986b). In each of these studies, work groups encountered and coped with new technological problems that were distributed widely over time and space. In Howard and Schneider's (1988) research, new problems emerged and new knowledge was created at various points in a linear production process. These points were located at the intersection of normal production problems (typically handled by workers through informal means prior to MRP II) and MRP II system deficiencies (which were caused by an absence of informal knowledge). Although Howard and Schneider (1988) do not address this issue directly, it is presumed that different individuals and/or informal work groups were involved in problem-solving sessions, depending upon the exact time and space location of the intersection points described above. Further, it is suggested (and this is an hypothesis warranting further investigation) that the sharing of new knowledge created at various time-space localities would be affected by the structure of formal and informal organization at the aircraft instruments plant (as indicated in Figure 1).

Orr's (1986a, 1986b) work does address the issue of distributed knowledge more directly, but in his case the distributed character of new knowledge is an obvious function of the geographical deployment of Xerox equipment. Orr's service technicians traveled in small teams, and these teams were the work groups that encountered new technological problems at geographically dispersed field sites. The teams altered their own mental models of machine operation by trial-and-error diagnostic and repair procedures in the field (which, interestingly, drew on distributed sources of information, including end-user reports; see Orr, 1988). New knowledge gained through dispersed trial-and-error problem-solving procedures was stored in team members' memories and shared through the communal exchange of narratives (which also, presumably, would be a process influenced by formal and informal organizational structure).

A final subsidiary proposition, one that also derives from our general theoretical stance and from Primary Proposition 2, concerns the content of local knowledge at advanced technology work sites.

> **Subsidiary Proposition 3.** Local knowledge systems at advanced technology work sites will include a new type of knowledge content, designated here as knowledge about systems.

Knowledge about systems (a phrase borrowed and adapted from Howard & Schneider (1988), whose meaning was more limited) is defined here as the informal conceptualization of relationships between different parts of large-scale technological processes, including parts that the individual has not experienced directly. This type of knowledge contains informal information about the ways that various parts of a total system do (or do not) fit together. Such knowledge may be used to enable workers to solve, predict, plan for or avoid problems that could occur at other times and/ or other places.

Support for the presence of knowledge about systems at advanced technology work sites is presented by Howard and Schneider (1988) and Sachs (1988). In both of these studies, informal work groups developed new bodies of knowledge which were used to *integrate* formal and informal understandings about real production processes on the one hand, and abstract MRP system rules and deficiencies on the other. Sachs (1988) presents an interesting example of knowledge about systems drawn from her study of MRP implementation at an electronics plant. At the plant, inventory control workers discovered a discrepancy between the number of parts of a certain type logged in MRP and the actual number of such parts found in a parts bin. In an effort to diagnose and correct this discrepancy, workers had to compare MRP's record of the parts' movements over time with the logic of actual production steps that were known (by the workers) to involve those same parts. This comparison enabled the workers to pinpoint the source of the discrepancy (i.e., a data entry error) and to reject their initial (and incorrect) hypothesis concerning the origins of the error. Subsequently, the workers were able to locate the missing parts and thereby complete a high priority job. This illustration, plus others presented by Howard and Schneider (1988), shows that informal bodies of knowledge contain information about relationships between different parts of complex technological processes, and that such information may be used to correct problems that originated at other places and times.

DISCUSSION

Kusterer's (1978) work was founded on the belief that workers carry knowledge paradigms as part of their cognitive apparatus, and that such paradigms are altered and expanded through the actual practice of work. While workers at traditional work sites clearly hold and use informal bodies of knowledge that may be described as paradigms (i.e., models or abstract representations of reality), it is the advanced technology worker whose knowledge resembles most closely the dynamic scientific paradigms envisioned by Kuhn (1970). The knowledge constructs of workers at advanced technology work sites are, like scientific paradigms, open systems whose content is constantly altered and expanded through encounters with new problems. And, like scientists, these workers continuously create and share new knowledge that pushes forward the frontiers of their paradigms in step with technological advance. Bodies of local knowledge at traditional work sites, on the other hand, may be more static over time and involve problem-solving practices which rely more generally on known procedures and techniques. It is possible that the dynamic problem-solving process engaged in by workers at advanced technology sites may be fundamentally different from the more static approach taken at traditional sites. Dynamic problem-solving may require more intensive collaboration among informal work groups, or may require substantial opportunities for experimental learning and/or knowledge dissemination among informal networks. It is also possible that differences in the problem-solving process engaged in by workers at traditional and advanced technology work sites may reflect differences in skill levels (although work groups studied by Howard & Schneider [1988] and Sachs [1988] probably were comparable in skill to workers at Kusterer's [1978] traditional site). Each of the possibilities noted above have clear implications for the management of advanced technology firms, and should be investigated through additional empirical research.

Our present understanding of local knowledge systems in advanced technology organizations presents another intriguing possibility, one that also has implications for management and suggests directions for future inquiry. Specifically, it is possible that bodies of local knowledge may be analogous to expert systems. Similarities between these two types of knowledge-based systems becomes clear when we examine the content of each.

An expert's knowledge is founded on a formal domain of data and theory, but also includes content derived from experience with a given class of problems. An especially important component of expert knowledge is the set of explicit and implicit means of evaluating concrete situations and forming plans of attack on complex problems (see Buchanan, 1982). The expert's problem-solving knowledge expands over time as the expert interacts with new types of problems.

Like expert systems, local knowledge systems at advanced technology work sites may be conceived as general problem-solving strategies that emerge from experience with a given class of problems. They also are founded on a formal domain of technical knowledge, include both explicit and implicit means of evaluating and approaching concrete situations, and expand over time in relation to new experience. Significantly, however, local knowledge systems at advanced technology work sites may be more complex than the knowledge carried by an individual expert due to their distributed and collective character; that is, the total content of local knowledge is dispersed over time and space among a group of worker-experts rather than being localized in a single mind. Indeed, it is known that the knowledge carried by individual experts (e.g., physicians, laboratory scientists) also varies across an expert population, and some of these populations have devised social means to pool their expertise (e.g., consultation, delphi techniques) for triangulation on especially difficult or important problems (Barr & Feigenbaum, 1982). What is not known is whether the structure and content of expert knowledge, aggregated over a population of experts, is analogous to the structure and content of local knowledge systems carried by informal work groups at advanced technology work sites. An answer to this question must await further progress in the field of artificial intelligence, as well as additional empirical evidence on systems of local knowledge.

If local knowledge is at least partially analogous to expert knowledge, then it may be possible ultimately to elicit and formally model local knowledge for inclusion in various types of automated systems. The general utility of such formalizations would, of course, depend upon the extent to which local knowledge is specific to a particular technology (versus specific to a particular organization), a question that also remains to be answered in future inquiry. The current problems involved in developing expert systems—including problems involved in using multiple individual experts—suggest that considerable time may pass before it is possible to formalize the collective and distributed expertise of work groups (see Barr & Feigenbaum, 1982). In the meantime, additional research could be undertaken to develop a better understanding of local knowledge systems and optimize their utility in advanced technology organizations.

The diagram presented in Figure 1 suggests a number of possible directions for future research. Several of the relationships depicted in the diagram are not well understood, including the interaction between local knowledge and organizational structure (both formal and informal), as well as the long-term feedback effects from local knowledge to formal technology. One of the most important relationships portrayed in the diagram, and one probably deserving our most immediate attention, is the dynamic interaction of formal technology and informal organization that creates local knowledge over time. This dynamic interaction (indicated by heavy broken lines in Figure 1) is especially significant because of its role in the process of technological innovation. Insofar as local knowledge supports the implementation of new technological processes and products (as suggested by the ethnographic literature), its creation may be viewed as part of the process of technological innovation (see Tornatzky et al, 1982; Tornatzky & Fleischer, 1990). From this standpoint, a better understanding of the knowledge creation process is crucial to the theory and practice of advanced technology management. Some important research questions in the area of local knowledge creation include the following: Exactly how do informal work groups interact with technology to create new knowledge at advanced technology work sites? What are the key variables that affect this process? Is informal collaboration across work groups required in all cases? Do informal leaders or especially gifted workers play a key role? Is new knowledge generation optimized when workers have a better understanding of formal technology? Do quality circles, sociotechnical job design or other types of organizational innovations enhance the creation of new knowledge?

Since technological innovation includes the process of knowledge dissemination, we also need more research on the ways and means by which new informal knowledge is shared among workers. Key research questions in this area include the following: What are the patterns of new knowledge flow and the social mechanisms that support such flow? Does the structure of formal and/or informal organization affect the dissemination of new knowledge across the work group? Do some individuals or informal groups withhold new knowledge from others? What is the rate of new knowledge dissemination and the extent of individual variability in access to new knowledge? Answers to these questions and others given above assume a continuing interest in informal organization, one that could lead us to rethink the position of informal work groups in advanced technology firms. Perhaps such work groups function in a manner that actively supports productivity objectives, more in keeping with the dynamic role envisioned by sociotechnical systems theory (Trist, 1981; Vavarek, 1987) and less like the static and restrictive role portrayed by the neoclassical literature (see Albrecht & Goldman, 1985).

Until the time that local knowledge can be elicited by knowledge acquisition procedures and formally modeled for inclusion in expert systems, managers are advised to map the localities of informal knowledge fields and preserve their integrity in the face of organizational and/or technological change. Because we do not yet have a clear understanding of the exact relationship between local knowledge and organizational structure, any projected change that alters the interface between formal technology and informal organization should be approached with caution.

Ethnographic research suggests that disruption of local knowledge fields (sometimes caused by severing the link between formal technology and informal organization) can cause serious production problems and lead to productivity decline (Howard & Schneider, 1988). Based on our current state of knowledge, work groups that interact with technology in the shop floor or office environment should be viewed as a collective group of experts, experts who must be consulted prior to and throughout the technological change process. Further, informal work groups should be granted adequate time and space to interact with technology in an experimental learning mode prior to full implementation of new technological processes.

Managers may expect to find complex and evolving systems of local knowledge anywhere that work groups interact with rapidly changing material or social technology. Orr (1986a) gives us an excellent means to detect the presence of expanding local knowledge fields; that is, wherever informal work groups share narratives (i.e., war stories) with a technical content. It should be noted, however, that narratives are only one way in which local knowledge can be packaged. Informal technical information also may be created and packaged in shop floor conferences, in the ad hoc training sessions that senior workers give to their juniors, or in organizational myths and ethnohistories. Such informal social phenomena, often portrayed as symbols of organizational culture (e.g., Martin, 1982), also may serve as flags that mark the location of a valuable technological asset. This asset, while virtually invisible on the surface of an organization, is a critical part of the technology-structure interface and should be recognized in future formulations of organizational theory.

NOTES

1. Culture also has been defined as a system of knowledge that is implicit and shared (see Goodenough, 1956). Culture in its traditional sense, however, is an holistic construct that refers to an encyclopedic body of knowledge shared by a social group (Werner & Schoepfle, 1986). In this paper I am using the concept of local knowledge system in a much more focused way (i.e., the informal information shared by a work group which contributes to the operation of a technological system).
2. Kusterer gives many vivid illustrations of machine operators' informal knowledge in each of the four categories listed here. He also presents corroborative evidence from a second case study of informal knowledge acquired by bank tellers (although tellers appear to have a higher ratio of formal to informal knowledge than machine operators). Finally, Kusterer describes more briefly a range of informal techniques and procedures acquired by workers in several different blue and white collar occupations.
3. In Japanese manufacturing organizations, informal knowledge about, and adjustments of machines are a standard feature of the on-going innovation process known as "giving wisdom to the machines" (Shimada & MacDuffie, 1990).

REFERENCES

Albrecht, S.L., & Goldman, P. (1985). Men, women, and informal organization in manufacturing. *Sociological Focus, 18*(4), 279-288.

Balkin, D.B., & Gomez-Mejia, L.R. (1984). Determinants of R&D compensation in the high tech industry. *Personnel Psychology, 37*, 635-650.

Barr, A., & Feigenbaum, E.A. (Eds.) (1982). *The handbook of artificial intelligence.* Vol. II. Los Altos, CA: William Kaufmann.

Blau, P.M., & Scott, W.R. (1962). *Formal organizations: A comparative approach.* San Francisco, CA: Chandler.

Braverman, H. (1974). *Labor and monopoly capital.* New York: Monthly Review Press.

Buchanan, B.G. (1982). New research on expert systems. *Proceedings of the Machine Intelligence Workshop, 10*, 269-299.

Burawoy, M. (1979). *Manufacturing consent: Changes in the labor process under capitalism.* Chicago: University of Chicago Press.

Doutt, J.T. (1959). Management must manage the informal groups too. *Advanced Management, 24*, 26-28.

Fry, L.W. (1982). Technology-structure research: Three critical issues. *Academy of Management Journal, 25*(3), 532-552.

Goodenough, W.H. (1956). Cultural anthropology and linguistics. *Bulletin of the Philadelphia Anthropological Society, 9*(3), 3-7.

Gouldner, A.W. (1954). *Patterns of industrial bureaucracy.* New York: The Free Press.

Homans, G.C. (1953). Status among clerical workers. *Human Organization, 12*(1), 5-10.

Howard, R., & Schneider, L. (1988). Technological change as a social process: A case study of office automation in a manufacturing plant. *Central Issues in Anthropology*, 7(2), 79-84.

Jewell, L.N., & Reitz, H.J. (1981). *Group effectiveness in organizations.* Glenview, IL: Scott, Foresman.

Kuhn, T.S. (1970). *The structure of scientific revolutions* (2nd ed.). Chicago: University of Chicago Press.

Kusterer, K.C. (1978). *Know-how on the job: The important working knowledge of "unskilled" workers.* Boulder, CO: Westview Press.

Lamphere, L. (1979). Fighting the piece-rate system: New dimensions of an old struggle in the apparel industry. In A. Zimbalist (Ed.), *Case studies on the labor process* (pp. 257-276). New York: Monthly Review Press.

Martin, J. (1982). Stories and scripts in organizational settings. In A.H. Hastorf & A.M. Isen (Eds.), *Cognitive social psychology* (pp. 255-305). New York: Elsevier/North Holland.

Orr, J. (1986a December). Narratives at work: Story telling as cooperative diagnostic activity. *Proceedings of the conference on computer supported cooperative work, Austin, Texas* (pp. 1-11).

________(1986b). *Talking about machines.* Unpublished manuscript, Intelligent Systems Laboratory, Xerox Palo Alto Research Center.

________(1988, November). *Work, practice, and community knowledge.* Paper presented at the 1988 Annual Meeting of the American Anthropological Association, Phoenix, Arizona.

Roethlisberger, F.J., & Dickson, W.J. (1939). *Management and the worker.* Cambridge, MA: Harvard University Press.

Roy, D. (1952). Quota restriction and goldbricking in a machine shop. *American Journal of Sociology, 57*, 427-452.

________(1954). Efficiency and the fix: Informal intergroup relations in a machine shop. *American Journal of Sociology, 60*, 255-266.

________(1958). Banana time: Job satisfaction and informal interaction. *Human Organization,*

18, 158-168.
Sachs, P. (1988, November). *Working knowledge.* Paper presented at the 1988 Annual Meeting of the American Anthropological Association, Phoenix, Arizona.
Shapiro-Perl, N. (1979). The piece-rate: Class struggle on the shop floor. Evidence from the costume jewelry industry in Providence, Rhode Island. In A. Zimbalist, (Ed.), *Case studies on the labor process,* (pp. 277-298). New York: Monthly Review Process.
Shimada, H., & MacDuffie, J.P. (1986). *Industrial relations and "humanware."* Unpublished manuscript.
Tornatzky, L.G., Eveland, J.D., Boylan, M.G., Hetzner, W.A., Johnson, E.C., Roitman, D., & Schneider, J. (1983). *The process of technological innovation: Reviewing the literature.* Washington, DC: National Science Foundation.
Tornatzky, L.G., & Fleischer, M. (1990). *The process of technological innovation.* Lexington, MA: Lexington Books.
Trist, E. (1981). *The evolution of socio-technical systems.* (Occasional Paper No. 2), Toronto Ontario Quality of Working Life Center, Ontario Ministry of Labor.
Vaverek, K. (1987). *The nature of semiautonomous'work group structure: An integration of the sociotechnical systems approach and group development theory.* Unpublished doctoral dissertation, University of Florida.
Werner, O., & Schoepfle, M. (1986). *Systematic fieldwork.* Vol I. Newbury Park, CA: Sage.
Whyte, W.F. (1948). *Human relations in the restaurant industry.* New York: McGraw Hill.
Wynn, E.H. (1982). The user as a representation issue in the U.S. In U. Briefs, C. Ciborra, & L. Schneider (Eds.), *Systems design for, with, and by the users* (pp. 349-358). Amsterdam: North Holland.

ORGANIZATION DESIGN:

A CRITICAL FACTOR IN HIGH TECHNOLOGY MANUFACTURING STRATEGY

Jan Zahrly

The traditional image of a manufacturing facility is based on a model of low cost production. Production methods have a highly rationalized work flow, have standard procedures and processes, narrow task assignments and routine inflexible activities (Mintzberg, 1983). This efficiency provides a competitive edge in the manufacture of high volume/high standardization commodity products (Hayes & Wheelwright, 1984). However, an organization design based on a model of efficiency is not adequate or appropriate for the modern high technology manufacturing facility. High technology factories have complex equipment and manufacturing processes, require highly skilled employees, demand flexible structures and operate in a dynamic environment. These characteristics of high technology manufacturing are not consistent with a traditional mechanistic organization.

Manufacturing strategy is the pattern of "structural and infrastructural decisions" over time that define how the product will be created (Hayes & Wheelwright, 1984). This functional strategy supports the organizational strategy and specifies how product or service creation will aid in the achievement of business unit goals (Wheelwright, 1984). There are multiple dimensions of manufacturing strategy including factors such as measurement and control systems, workforce policies, and organizational structures.

Traditional manufacturing strategies are usually understood and accepted by employees. In order to implement less traditional manufacturing strategies that support high technology production processes, the manager must provide education and leadership to bring about awareness and acceptance of the new organization design. The implementation of high technology manufacturing strategy is time consuming, expensive and must include substantive and symbolic components.

HIGH TECHNOLOGY MANUFACTURING

The strategy for high technology manufacturing demands a flexible organization. Complex equipment, highly skilled employees and a dynamic external environment propel the organization design toward an adaptable structure that facilitates innovation and rapid change (Burns & Stalker, 1961; Davis & Lawrence, 1977; Thompson, 1964). Such an organization design is costly in that it requires many resources for training, recruitment and orientation. Highly skilled employees are usually mobile and have little loyalty to the organization (Mills, 1985), leading to high turnover costs. A flexible organization design is not consistent with the "traditional" goal of efficiency that is based on a minimum of slack resources. Because the skilled labor force is so mobile, the manager must give attention to those organizational characteristics that will enhance the employees' perceptions of the workplace and that aid in the retention of employees (Lawler, 1988; Mills, 1985). Employee expectations and a dominant external environment influence the manager of a high technology manufacturing firm to provide the flexibility and resources for frequent innovations and modifications. In addition, the manager must foster an internal environment that tolerates flexibility and change.

The level of technological complexity influences the extent of mechanization and, therefore, the predictability of the production process (Woodward, 1965). In the past, high technology organizations tended to be highly dependent on machines and equipment while low technology organizations utilized much human labor in the production process. Contemporary high technology manufacturing organizations frequently utilize continuous process production methods with a high degree of automation. These factories are often quite dependent upon complex human labor. Keller (1978) demonstrated that the performance of continuous process organizations was positively related to organic organizational structures (i.e., flexible, less routine, higher employee participation in decisions, and so on). This indicates a need for organic structures in high technology manufacturing. In recent years, building upon the Woodward (1965) typology of technical complexity, high technology has also come to represent complex and intricate human processes. Highly complex technical equipment creates a demand for highly skilled designers, operators and repairpersons.

Standard procedures and routine technology, with low variety and high analyzability, are economically successful in mass-production firms (Woodward, 1965). The routinization of automated, mechanistic tasks constrains the degree of flexibility available in determining the organization structure. However, highly technical firms involved in mass production often encounter high variety, low analyzability or nonroutine technology. This is a result of the dynamic and changing environment and of the rapid development of new machine technology. While routine standardized activities are profitable manufacturing techniques and routinization is the basis of stable organizations in the long run (Trice & Beyer, 1986; Weber, 1947), routine activities are usually not adequate for high technology manufacturing. Nonroutine actions and flexible situations are necessary in order to cope with rapid change. The manager of a highly technical manufacturing firm faces a dilemma of how much flexibility to build into the system. On the one hand, the manager prefers unchanging, routine, standard activities that are cost efficient and profitable. On the other hand, the dynamic environment forces a flexible manufacturing strategy in order to remain competitive. One way of dealing with this dilemma is to provide symbolic flexibility. The successful manager can use symbols, concomitant with substantive structural components, to demonstrate desired values that aid in goal accomplishment.

SYMBOLISM IN ORGANIZATIONS

An emerging cognitive perspective on organizations is that reality is socially constructed, that is, people create their own realities based upon their own histories, personalities, value systems, and so forth (Astley, 1985: Daft & Weick, 1984; Pfeffer, 1981). Enactment of reality is the process of giving meaning to various objects and events in the environment. Because people interpret reality in differing ways based upon different experiences, it is important for managers to provide a common meaning within the organization. By providing meaning, managers present more opportunities for employees to interpret words and actions (Daft, 1983).

Symbolism is one method of providing meaning. A symbol may be a process or an isolated event. It may be an action or an inanimate object; symbols may be explicit or implicit. A symbol is anything that is used to represent something else. The primary function of symbolism for humans is to facilitate communication of meaning (Eoyang, 1983). Symbolic forms are an integral part of any culture and, as such, should be analyzed in their specific organizational context.

Managers and leaders must engage in symbolic action. Smircich and Morgan (1982, p. 258) assert that "leadership is . . . the process whereby one or more individuals succeeds in attempting to frame and define the reality of others." Pfeffer (1981, p. 5) argues that "management's effect is primarily with respect to expressive or symbolic actions," while Mintzberg (1973) notes that one of the roles of a manager is to be a figurehead, a symbol. Others have created similar terms of impression management (Goffman, 1959) and dramaturgy (Thompson, 1977). Effective managers will use symbols to create a reality that is understood by all employees and allows the realization of organization goals.

This chapter presents examples of symbols that aided in the definition of the manufacturing strategy of a high tech factory start-up. The plant was designed, built and operated by a subsidiary of a Fortune 500 corporation. Qualitative and quantitative data were gathered over a two year period. Modes of data collection included survey questionnaires, interviews (structured and unstructured) and participant observation.

The research occurred in a small Southern town, the factory was a green-field site, hourly employees were new to the industry and there was no union. There were 80 hourly employees and 15 managers/support staff. Many of the managers were from other factories in the industry and knew the basic production process. However, all of the equipment was prototypical, designed by the company's engineers in conjunction with suppliers, and the organization structure was unusual in the industry. Teamwork and high employee participation were planned. The production tasks and equipment were complex; the production process was highly automated and integrated. Production demanded highly skilled employees for operation, repair and modification of equipment, as well as analysis of complicated problems. The work schedule was physically demanding while the tasks were mentally demanding.

The researchers were invited into the facility to observe and investigate the development of the sociotechnical system in the organization. The author observed each team at the factory site at least once every six working days. She was given unlimited access to all areas of the facility, participated in social activities with employees away from the site, and occasionally performed unskilled tasks on the production floor.

SYMBOLS AND MANUFACTURING STRATEGY

Many symbolic acts and artifacts were observed during the two years of the research project. The data presented here describe actions that were effective in defining manufacturing strategy reality for the workforce. Four dimensions of manufacturing strategy were examined: organization structure, selection and socialization processes, communication activities and patterns, and reward systems. The manufacturing strategy of this firm included many components, such as standard health insurance, which would probably be found in most manufacturing facilities. This discussion revolves around those design components and symbols that would not be found in a traditional manufacturing firm. Table 1 provides examples of each of these strategic dimensions.

Structure

The organizational structure was designed to facilitate the plant start-up with a minimum of supervision. Semi-autonomous permanent work teams were created and worked together to solve problems. Jobs were rotated among team members; this required cross training for all tasks. No one "owned" a job and task assignments were exchanged based upon ability, training and production demands. The quality assurance jobs were also rotated among all of the employees. This rotation guaranteed that all team members would have a better understanding and commitment to high quality production. There was no paid sick leave and no substitute employees were hired when a worker was absent. The team completed the necessary tasks to maintain production, regardless of absences. There were twenty people on each team; an hourly worker served in a semimanagerial capacity, assisting the team superintendent and performing supervisory and organizational tasks. This position was also rotated over time.

Most ratings and evaluations were on a team basis. Daily production, quality ratings, repair work and preventive maintenance were reported and/or assigned by team. Team members developed great loyalty to their teams and a competitive attitude toward other teams. When some team members were not getting breaks at regular times (and complained about the lack of breaks), managers and team superintendents refused to draw up a break schedule. They told the teams to work out the breaks among themselves. The result was that each team designed a rotating break schedule to assure equal breaks for all employees (different teams designed different schedules). Managers were vocal in praise of the team scheduling and encouraged the teams to continue developing ideas to improve the work place. Employees took the encouragement to heart and reworked holiday and vacation schedules among themselves. Entire teams even exchanged shifts to improve the holiday schedules of everyone.

One demonstration of flexibility in highly technological organizations may be low formal structure (Lawrence & Lorsch, 1969). There were only two levels of supervision in the entire factory. The team supervisors and other technical staff members reported directly to the plant manager. All hourly employees and support personnel reported to first level supervisors.

Table 1. Components of High Technology Manufacturing Strategy

Structure

Permanent work teams

- No substitute employees—team did all the work when a team member was absent—no paid sick leave
- Teams scheduled breaks among themselves
- Teams reworked vacation schedules among themselves
- Cross training on all tasks
- All tasks were rotated, including quality assurance

Ratings, rewards based on team and plant performance
Hourly employee in semi-management position on team
Ombudsperson
Low formal structure

Selection and Socialization

Many pre-employment interviews by managers
Job offer to applicant and spouse/partner by plant manager
Expectations were repeated often during pre-employment interviews - created desired ideology
Selection criteria on skills and desired personal characteristics (ability to work in a team, self-starter, high work orientation, and so forth)
Training in-house for groups and individuals
Most employees trained other employees
Initial group orientation and training

Communication

Monthly paid team meetings with plant manager
Informal daily team meetings encouraged
Annual plant-wide communications meeting with corporate executives
Informal lines and methods of communication
Language patterns
Plural pronouns to include everyone
Use of first names for everyone

Reward Systems

Compensation

- Skill based pay
- Extra vacation day if employee worked one year with no absence or tardiness
- Quarterly bonus based on quality and quantity of production, safety, and attendance

No time clock
Free local phone in break room
Team evaluations, ratings
One break room/cafeteria for all employees
No reserved parking spaces

An ombudsperson was employed and had an office near the plant floor, beside the quality assurance lab. The person who held this position reported to the plant manager; his background was in human resource management and he perceived his task as liaison between management and labor. Hourly employees used him as a resource person for information and as a counselor for problems. Counselling by the ombudsperson was in addition to a standard EAP program in the organization.

All of these structural components were symbols of management's belief that employees could manage themselves. The team concept, with hourly workers in semimanagement positions, reinforced the respect for employees. By encouraging cross training and task rotation, managers demonstrated their belief that each employee was capable of doing all tasks. This was flattering and encouraged employees to strive for additional training. Employees were praised when they attained a higher level of skill and when they were able to accomplish additional tasks. This led to a sense of personal responsibility on the part of most employees. Each employee believed that he or she personally contributed to the production efforts. However, the individual was almost always subordinated to the team. The team ratings, the design of a break and holiday schedule by the teams, the job rotation were all symbols that the team, rather than the individual, was most important. Because the symbols were there, reinforced by management, hourly workers enacted team management and learned to accomplish the job, often with no direct supervision. Pride in the team's quality and production was evident and was reinforced by public postings of team ratings on bulletin boards around the plant. The perceptions of personal responsibility and team self-management became a reality to these employees.

Selection and Socialization

The top managers at the factory knew that the work shift was arduous (12 hours on and 12 hours off, three days on and three days off) so they designed a hiring procedure that would aid in the selection of highly qualified employees as well as assist in gaining compliance and setting the ideology of the organization. Because the work shift was incompatible with the standard work week, there was a potential for severe conflict between work and family/social roles. After an applicant passed the preliminary skill screening, he or she attended a company designed 42 hour course at the local community college. There was no charge for the course but applicants had to complete the course on their own time. Each applicant then had a minimum of eight interviews with managers. Interviews lasted about one hour each and followed no protocol. However, managers did have an unwritten set of desired employee characteristics (e.g., ability to work without supervision, strong work ethic, good communication skills, etc.) that they believed would contribute to high participation and team effort.

Final selection was by management consensus and the applicant and his or her spouse/partner were invited to meet the plant manager, who made the offer of employment to both people while they were seated in the manager's spacious office. The plant manager carefully orchestrated the hiring ritual; he explained that the company was made up of team players and that the new employee and the spouse/partner would both be part of the team. Furthermore, it was explained to the spouse/partner that the work schedule was demanding but the spouse/partner was expected to be understanding. From preemployment through the first few months of employment, management was co-opting employees and their family members, seeking their symbolic participation by obtaining implicit agreement to the terms of employment, including the strenuous schedule. The plant manager later confided to the researcher that he made the offer to the couple, in his office, because he knew that the office was impressive and he hoped to gain worker and family acceptance of the work schedule by personally presenting the employment offer.

This hiring ritual was a symbol of the importance of the employee and the employee's family to the company. Trice (1985, p. 227) points out that rituals are "directed at managing anxiety" and the plant manager used the hiring ritual to reduce or manage the anxiety of the new employee and his or her spouse/partner. Even though the factory job was as demanding and demeaning as any factory job, and possibly more demanding because of the difficult schedule, the plant manager used the symbolic cooptation actions to forestall complaints and to gain compliance in the short run.

Management also worked at establishing the organizational ideology during the preemployment period by constantly telling the applicants, during the interview process, about the values that were deemed to be important for success in the company. Applicants were told that they would be permanently assigned to semi-autonomous teams and would be expected to work with the team and participate in many decisions. In fact, there were very few decisions that the team could make since the equipment was so highly automated and integrated. Operatives could design modifications to equipment but could not implement the changes until they were approved by management. Operatives could not change tasks or work schedules unless approved by management. Nonetheless, most operatives believed they worked in a participative workplace and they believed that they had decision-making power over many aspects of their jobs. Top management did consult with employees when feasible and implemented many employee decisions. Managers also forced teams to make decisions when it was possible for teams to design and implement procedures and rules (e.g., schedules for breaks). While these operatives were not involved in participation in the traditional sense of the term (Ouchi, 1981; Vroom & Yetton, 1973), they participated in more decisions than most of them had previously experienced in other jobs. In the meantime, management continued the difficult task of getting a new plant operational. Only after the plant was fully operational did most operatives realize that their influence over decisions had been minimal. By inculcating the myth of participative management during hiring interviews, and by reinforcing participation by forcing teamwork and team decisions, the managers provided a symbolic flexible structure. At the same time, operatives were redefining the meaning of participative management. While the level of decision making by employees remained low in many situations, the employees had a sense that they could influence decisions about the work environment and the tasks.

Employees initially entered the organization in groups. They received orientation and training as a group and developed cohort alliances which lasted for many months. However, pressure for production led managers to change the orientation mode so that half of the employees entered the organization individually and learned on the job. Later research findings demonstrated that employees who experienced group orientation and training were more satisfied with the job and had lower work-family stress than employees who experienced individual training and orientation (Zahrly & Tosi, 1989). At the end of the study, managers expressed regret that all employees had not received formal group training and orientation because they believed group orientation led to greater company loyalty.

All training, except for early equipment training at a vendor's site, occurred at the factory. A specific location was established on the premises for training and for developing training materials. As employees became more knowledgeable about the equipment and processes, they trained others. Employee training occurred individually and in groups for the first two years of operation. Skilled employees led formal classes and seminars for their fellow workers. As soon as employees mastered various skills, most of them trained other employees in the same tasks. All employees trained others individually on the plant floor at one time or another. The attitudes toward self-training reinforced the management concept of participation by all employees. It also created a confidence that these employees could manage themselves and were able to do what was necessary, including teaching others, to achieve production and quality goals.

There is no doubt that the lengthy selection process and the comprehensive cross-training were costly in financial and human resources. The human resources manager explained that "soft" dollars spent on selection, training and liaison persons such as the ombudsperson could rarely be defended in financial terms. However, the selection and socialization processes initially established the value systems of teamwork, self-training and personal cohort support. Symbolic acts of managers reinforced these values until they were institutionalized by the employees.

Communication

Team meetings between all team members and the plant manager were held monthly during the first two years of operation. These meetings were scheduled and employees were paid overtime to attend the meetings. They were called "communication meetings," conducted by the plant manager who was very emotional and expressive in verbal presentations, and he did most of the talking in the meetings. Production numbers, quality ratings and modifications to equipment were the primary topics. However, the plant manager always stressed that the meetings were for "two-way" communication, he used plural nouns and pronouns such as "we, our, us" to convey that management and operatives were all involved, he talked of having an "open door" to all employees, and, in general, used language to convince employees that they were decision makers in the organization. In addition, managers insisted that all employees use first names for everyone. This reinforced the sense of egalitarianism that the plant manager wanted to foster.

The monthly meetings between teams and the plant manager served another purpose. Management could reduce the formal communication structure, yet maintain the top-down communication. Much information was communicated in the informal monthly meetings. In addition, informal daily team meetings occurred where the team supervisor could transfer important information to the employees. Employees were not paid for the time in the informal daily meetings but attendance was noted and encouraged by supervisors and informal team leaders. Attendance soon became the norm and nonattendance was punished by withholding information and obvious verbal harassment by peers. Managers were aware that nonattendees were harassed and did not intervene. The pattern was primarily top-down communication and almost no bottom-up communication. However, since the communication network was so informal the employees felt that the structure of the entire organization was informal and this gave them flexibility. Every employee had frequent opportunities to communicate with all managers. In fact, there was little flexibility in the production process, many tasks were precisely defined and the ordering of tasks was exact, with no deviation. Nonetheless, the language of the managers and the informal communication network allowed the managers and employees to insist that flexibility existed when it did not.

An annual plant-wide meeting with corporate executives demonstrated the organization's communication patterns. Once a year the factory was closed and all employees and spouse/partners were invited to a meeting away from the factory. These were breakfast or dinner meetings and were held at a private golf club. Executives made short speeches and then responded to questions from the employees. As the employees left the meeting, they received gifts (e.g., mugs, tee-shirts, caps, and so on) with the company logo prominently displayed. Most of the employees were very impressed with the scope and grandeur of the annual meeting. Many of them had never been in a golf clubhouse or met corporate officers before. In general, they believed they were important to the company (because corporate officials had come to give them information) and the annual meeting was a demonstration of the value of operatives.

Communication patterns reinforced three facets of the organization's ideology. The first was the informality of all communication. Written directions or information were rare and managers were easily accessible to all employees. This informality constantly reminded the employees that the organization design was flexible and adaptive. The second dimension of communication patterns was the plurality. Individualism was never emphasized or encouraged. Team performance and team efforts were rewarded. Finally, the communication meetings demonstrated management's high regard for all employees. These meetings reinforced the organizational ideology and employees' belief in themselves.

Reward Systems

Compensation was based on skill levels, regardless of the actual task performed by the employee. There was no time clock; employees turned in a pay card once a week that listed the hours the employee had worked. Employees who were neither tardy nor absent for one year were given an extra paid day of vacation. The names of those who had earned the extra day were posted prominently in the break room. There was only one break room/cafeteria that was used by all employees, hourly and managerial.

The parking lot was well drained because the plant manager believed that employees should not have to get their feet wet when coming to or leaving work. There were no assigned spaces in the parking lot, with the exception of a space reserved for the plant manager. A free phone (restricted to local calls) was in the break room. Peer pressure prevented abuse of phone time. These are small rewards when recognized individually but they were symbolic of an egalitarian attitude around the factory. Employees believed that these symbols indicated that hourly workers were as important as managers and support staff. Their actual jobs were quite repetitive and boring after the initial excitement of the new job wore off and as the equipment became operational. Still, the symbolic message was that all employees were valued and all were treated equally in many respects.

Production and quality ratings were reported by team, and rewards were given to teams. For example, special meals were catered in for teams who achieved very high ratings in safety or production. There were almost no individual rewards—the sole exception being public praise for individuals who had developed new adaptations or had created particularly unique and helpful devices for the work flow. Employees were identified by team and developed great pride or shame in the team, usually based upon team production.

A quarterly bonus plan was instituted after the factory became fully operational. The bonus amount was based on quality and quantity of production, safety and attendance. Standards were team and plant-wide ratings. The quality and quantity of production was a function of team efforts while attendance and safety were primarily individual actions. If employees were lax about attendance or safety, peer pressure influenced individual behavior until the individual behavior improved. Team pride was a driving force influencing each team to high quality production. If any team failed to meet quality and quantity standards for a quarter, thereby depriving all hourly employees of bonus monies, the offending team was harassed and isolated. Standards for the bonus were clearly defined by management and were enforced by peer pressure.

The reward system was a symbolic representation of the desired values of teamwork and a sense of personal responsibility for the organizational goals. The rewards were designed to demonstrate that all employees would benefit if all employees would be responsible for their personal and team behavior.

IDEOLOGY AND CULTURE

There were many actions that symbolically demonstrated that the employees were being treated with respect. These actions reinforced the ideology of personal responsibility on the part of all employees and led to stronger attitudes of egalitarianism.

One of the best examples of the development of an ideology was the symbolic round table. The plant manager ordered a very large round table for the conference room at the facility. Operatives were rarely in the conference room; it was used mainly by corporate staff and managers. The table was so large that it had to be transported in pieces and put together in the conference room. After the table was installed, the plant manager insisted that every employee go into the conference room to see the round table. All visitors were taken to see the round table. The plant manager explained to all employees and visitors that the round table was a symbol of equality—no one sat at the head of the table because there was no head of the table; no one at the factory had priority or power over anyone else. Again, there was no doubt that the plant manager and all of the supervisors had tremendous power over the operatives. However, the inculcation of this myth of egalitarianism produced a litany that the researcher heard over and over during the first year of operation. The litany varied in exact text but the basic theme was the same: "We have a round table. We all participate in making decisions. We are all equal here." By the time the majority of the operatives realized that the "round table" myth was just that—a myth—the plant was operational and both operatives and managers alike used the "round table" image to represent various points of view. Those who wanted to continue the myth of egalitarianism spoke of the table in a positive manner. Employees who wanted to describe a failure to achieve complete participation used the table as an example of the failure of the system.

Symbolism was also utilized to demonstrate covert values. For example, the focus of goal emphasis was incongruous, particularly for operatives. Managers always spoke of having a primary goal of a high quality product. This is often the goal of high technology organizations (Perrow, 1970) and the operatives initially believed that management did value quality above other goals. However, the true goals of management soon became obvious through the actions of the plant manager. A few days after the plant became operational, daily listings of quantity of production were posted on the bulletin board in the break room/cafeteria. By the fourth or fifth month of operation, there were four different bulletin boards around the factory where production numbers were posted. Production was listed by day and by team. One board on the production floor listed the rate of production on a three hour basis and indicated the expected quantity of production at the end of the workday if production continued at the same pace. After one year of operation, a single bulletin board was placed near the entrance to the break room/cafeteria, which listed the monthly quality rankings of the product. This was the only visible record of quality performance at the facility. Therefore, while the plant manager often spoke of quality ratings and insisted that quality was the highest goal, the quantity of production was the number that was posted for everyone to see. Employees quickly understood the plant manager's priorities and placed quantity of production over quality of production. This is not meant to indicate that quality was not a goal of the organization. Quality goals were subordinated to quantity goals.

The key symbolic components of this structure were the establishment of teams, the frequent "communication" meetings between the plant manager and the teams, and constant verbal iteration that the plant operated under a participative management style. These elements provided the managers with a forum for shaping the reality of employees. Managers used the semi-autonomous teams for minor decision making, then praised the participation of employees in public meetings. All of this occurred after management established the "existence" of a participative management style during employment interviews.

The results of a high technology manufacturing strategy that includes symbolic actions indicate that such an effort was successful. Turnover, absenteeism and tardiness were almost zero for the first two years. Quality of the product was high and production and profit goals were exceeded every year. Teams worked on problems and shared information frequently. Innovation occurred, particularly in process and equipment adaptations.

The symbolic activity, concomitant with certain structural components, was successful. The managers were able to accomplish a difficult plant start-up, the employees were satisfied with their roles in the "participative" environment, and most of the employees remain with the organization five years after the start-up. Most employees eventually came to the realization that while the management style was not "participative" in the traditional sense, it afforded employees some influence in the quality of their work lives. By the time the workers came to their own realities, which often differed from management's realities, most of them were satisfied enough in the job to remain. The pay scale was high and only two out of 80 employees left during the first two years of operation. (One quit to return to college and one was fired for poor performance.) Surprise and disconfirmation did occur but it may not have been as severe as researchers believe (Louis, 1980). The symbolic gestures and actions on the part of managers assured employees of a flexible organization structure that was within their range of acceptance. While the manufacturing tasks were routine and resistant to change, the employees worked with and around the organization structure to enact a reality that allowed adaptation and innovation.

Practicing managers may draw many implications from the success of this effort. It is possible to instill an attitude of innovation and flexibility in a high technology manufacturing facility by the use of structural mechanisms and symbolic behavior. Only a few examples are presented in this chapter. Managers should search for symbolic actions as well as structural dimensions that will enhance teamwork as a high technology strategy. The isolated action or the single design component does not bring about an attitude of flexibility and change among high technology employees. It is the combination of many actions and design components, along with the frequent repetition of symbols, which leads employees to adopt common value systems.

A manufacturing strategy such as this is initially expensive. However, skilled employees are required for the operation and repair of the complex equipment and production processes. The costs of high turnover and absenteeism, frequent training of new employees, and more managerial staffing probably outweigh the costs of cross training and lengthy selection procedures. Additionally, there was no attempt to unionize the employees during the first two years of operation. This is a significant outcome since every other factory in the parent company was unionized. While it is easier to implement a total manufacturing strategy such as this in an organizational start-up, it is also possible to change existing strategies in an ongoing manufacturing facility by using some of the design components and symbols described in this chapter. Existing strategies may experience incrementally small changes over time that lead to significant changes in the long run.

Research should continue on the influence of manufacturing strategy on total organizational performance. Traditional manufacturing dimensions, when utilized in high technology organizations, are dysfunctional and expensive.

ACKNOWLEDGMENT

Data collection was supported by NSF grant #8314210ISI.

REFERENCES

Astley, W.G. (1985). Administrative science as socially constructed truth. *Administrative Science Quarterly, 30,* 497-513.

Burns, T., & Stalker, G.M. (1961). *The management of innovation.* London: Tavistock.

Daft, R.L. (1983). Symbols in organizations: A dual-content framework for analysis. In L.R. Pondy, P.J. Frost, G. Morgan & T.C. Dandridge (Eds.), *Organizational symbolism* (pp. 199-206). Greenwich, CT: JAI Press.

Daft, R.L., & Weick, K.E. (1984). Toward a model of organizations as interpretation systems. *Academy of Management Review, 9,* 284-295.

Davis, S.M., & Lawrence, P.R. (1977). *Matrix.* Reading, MA: Addison-Wesley.

Eoyang, C.K. (1983). Symbolic transformation of belief systems. In L.R. Pondy, P.J. Frost, G. Morgan & T.C. Dandridge (Eds.), *Organizational symbolism* (pp. 109-121). Greenwich, CT: JAI Press.

Goffman, E. (1959). *The presentation of self in everyday life.* Garden City, NY: Doubleday.

Hayes, R.H., & Wheelwright, S.C. (1984). *Restoring our competitive edge: Competing through manufacturing.* New York: Wiley.

Keller, R.T. (1978). Dimensions of management system and performance in continuous-process organizations. *Human Relations, 31,* 59-75.

Lawler, E.E. (1988). Choosing an involvement strategy. *Academy of Management Executive, 2,* 197-204.

Lawrence, P.R., & Lorsch, J.W. (1969). *Organization and environment.* Homewood, IL: Irwin.

Louis, M.R. (1980). Surprise and sense making: What newcomers experience in entering unfamiliar organizational settings. *Administrative Science Quarterly, 25,* 226-251.

Mills, D.Q. (1985). *The new competitors.* New York: Wiley.

Mintzberg, H. (1973). *The nature of managerial work.* New York: Harper & Row.

________(1983). *Structure in fives: Designing effective organizations.* Englewood Cliffs, NJ: Prentice-Hall.

Ouchi, W.G. (1981). *Theory Z: How American business can meet the Japanese challenge.* Reading, MA: Addison-Wesley.

Perrow, C. (1970). *Organizational analysis: A sociological approach.* Belmont, CA: Wadsworth.

Pfeffer, J. (1981). Management as symbolic action: The creation and maintenance of organizational paradigms. In L.L. Cummings & B.M. Staw (Eds.), *Research in organizational behavior* (Vol. 3, pp. 1-52). Greenwich, CT: JAI Press.

Smircich, L., & Morgan, G. (1982). Leadership: The management of meaning. *Journal of Applied Behavioral Science, 18,* 257-273.

Thompson, V.A. (1964). Bureaucracy and innovation. *Administrative Science Quarterly, 10,* 1-20.

________(1977). *Modern organization* (2nd ed.). University, AL: University of Alabama Press.

Trice, H.M. (1985). Rites and ceremonials in organizational cultures. In S.B. Bacharach & S.M. Mitchell (Eds.), *Research in the sociology of organizations* (Vol. 4, pp. 221-270). Greenwich, CT: JAI Press.

Trice, H.M., & Beyer, J.M. (1986). Charisma and its routinization in two social movement organizations. In B.M. Staw (Ed.), *Research in organizational behavior* (Vol. 8, pp. 113-164). Greenwich, CT: JAI Press.

Vroom, V.H., & Yetton, P.W. (1973). *Leadership and decision-making.* Pittsburgh, PA: University of Pittsburgh Press.

Weber, M. (1947). *The theory of social and economic organization.* New York: Oxford University Press.

Wheelwright, S.C. (1984). Manufacturing strategy: Defining the missing link. *Strategic Management Journal, 5,* 77-91.

Woodward, J. (1965). *Industrial organization: Theory and practice.* London: Oxford University Press.

Zahrly, J., & Tosi, H. (1989). The differential effect of organizational induction processes on early work role adjustment. *Journal of Organizational Behavior* (Vol. 10, pp. 59-74).

PART II

POLITICAL PROCESSES AND THE MANAGEMENT OF GROWTH IN HIGH TECHNOLOGY

THE POLITICS OF RADICAL TECHNICAL INNOVATIONS

Robert A. Page, Jr. and W. Gibb Dyer, Jr.

Innovation is not only a technological process, but a social and political process as well. A growing number of researchers have consistently called attention to the importance and impact of political processes on the development of new ideas and products (Dyer & Page, 1988; Kanter, 1983, 1988; Miller & Friesen, 1982; Normann, 1971; Sapolsky, 1972). The research that has been done typically focuses on the politics of "regular" innovative activity—incremental enhancements of established technologies. In this chapter we will focus on the processes involved in radical, discontinuous types of product and process innovation—the dramatic technological breakthroughs that reorient entire firms and industries (Abernathy & Clark, 1985; Abernathy & Utterback, 1977; Tushman & Romanelli, 1985). Radical innovations characteristically: (1) offer new technological designs fundamentally different from past designs; (2) demand new organizational systems, procedures and organization; (3) destroy the value of existing expertise; and (4) undermine the value of existing knowledge bases and open links to whole new scientific disciplines (Moore & Tushman, 1982; Perry & Page, 1986).

While the definition of politics remains a subject of considerable controversy, and the phenomenon itself is "messy" and "elusive" (Frost, 1988), for our purposes, we define organizational politics as "organizational activities intentionally designed to promote the interests of certain individuals or groups" (Frost, 1988; Gandz & Murray, 1980). This definition is meant to be interpreted broadly, under the premise that most organizational activities involve some degree of politics (Mayes & Allen, 1977). As Frost (1985, p. 504) notes, organizational activities typically include:

> Contests among interdependent actors . . . for control of resources, for the ability to determine the means/ends of doing organizational work [and] struggles for collaboration in the performance of work when the means/ends of its accomplishment are unclear and/or subject to dispute.

Indeed, politics and power relationships fuel both the development and resolution of these contests and struggles (Frost, 1988).

In this chapter we will describe the political processes involved in successful radical product innovation, focusing on how different political processes interfere or facilitate innovation, and on the role of multilevel interpersonal networks within such processes. While there has been relatively little research on this topic, what has been done presents a relatively consistent, albeit fragmented picture of the political nature of innovation. Beyond integrating this research, we will draw on data from two extensive, longitudinal studies of innovative companies to illustrate our analysis: Burgelman and Sayles, (1986) study of "United," and the Brigham Young University Innovation Group's study of "Multi-Graphics, Inc" (MGI) (Dyer & Page, 1988). Both studies supplemented quantitative data and archival information gathering with extensive qualitative research—primarily in-depth interviews. This qualitative data provides rich, contextual data critical for understanding informal processes such as organizational politics.

POLITICS AND RADICAL INNOVATION

We argue that radical product innovations tend to stimulate more political processes than other forms of innovation. Radical innovations hit at the core of political activity—they introduce high levels of ambiguity and uncertainty, they typically challenge vested interests, and, if pursued, they threaten to divert resources from other projects, and their constituencies (Kanter, 1988; Normann, 1971). A review of the literature suggests that radical innovations can be categorized into four major types: revolutionary, systematic, idiosyncratic and marginal. These types of radical innovation are depicted in Figure 1 and will be discussed according to the complexity and difficulty of the political and social processes necessary for their success. Revolutionary or "architectural" radical innovations involve technologies with potential so dramatic that they create their own markets

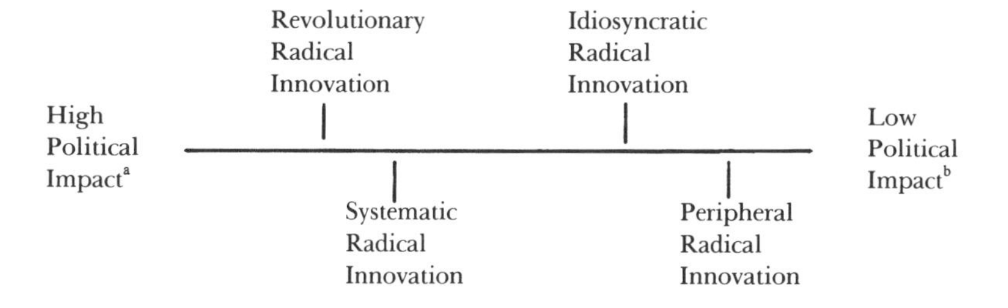

Figure 1. Types of Radical Innovation and Political Impact

Notes: [a]High Political Impact: Strategic Domain, Structure, and Political Relationship are affected and changed across all organizational units.
[b]Low Political Impact: Strategic Domain, Structure, and Political Relationship are unaffected and unchanged for most organizational units.

(Abernathy & Clark, 1985; Burgelman & Sayles, 1986; Dyer & Page, 1988). The organizational reorientation necessary to pursue such a major new product is so enormous that Tushman and Romanelli (1985) label it "re-creation"—all previous norms concerning strategy, structure and political relationships become precarious. Given the magnitude of change being introduced, this type of innovation is the most traumatic and involves the greatest degree of risk, for the product technology may not emerge as the dominant technological design in the new market (Miles & Snow, 1978). Because it is often vigorously resisted and its legitimacy is easily attacked, revolutionary innovation involves extensive political processes and negotiation.

Systematic radical product innovation involves developing the next generation of technology for existing or related markets. Consequently, while organizational structure and power relationships are often affected, this type of innovation falls into accepted definitions of the organization's strategic domain and context. While often difficult and traumatic, such reorientations are recognized as legitimate. They are periodically necessary to remain competitive as business environments and technologies evolve (Abernathy & Clark, 1985; Miles & Snow, 1978; Tushman & Romanelli, 1985).

Idiosyncratic radical product innovations involve a strong, central power figure, typically a key executive, who "is exceptionally strongly influenced by some event and then legitimizes and promotes a project" (Normann, 1971, p. 208). To the extent this type of project does not result in organizational "reorientation" around the emergent technologies, it becomes idiosyncratic. While changes in organizational values may be announced, they are seldom accompanied by corresponding changes in the formal structure and in power relationships (Normann, 1971).

Peripheral radical innovations involve relatively smaller projects that are of marginal importance because they do not involve major markets or major areas of research emphasis. While these projects often go outside of the organization's traditional strategic domain, because of their size they do not threaten existing organizational goals, values, structure or political arrangements, and are typically fitted into preexisting production and distribution systems (Normann, 1971). Predictably, the political and social processes needed to legitimize and sustain this type of innovation are less complicated because they minimally affect the status quo. While the technologies involved may undergo radical change, the organization does not.

Consequently, the process of radical innovation involves overcoming political as well as technical challenges. Developing a superior technology alone is no guarantee of success, for organizations have proven quite adept at suppressing radically innovative technologies, in some cases delaying their introduction into the industry for decades (Dunford, 1987; Frost & Egri, 1988). Political concerns need to be addressed, for "political" problems are the leading cause of failure of New Venture Departments in corporations (Fast, 1979).

To explore these issues in greater depth, we will examine the political processes typically involved in developing a radical innovation from initial conception through prototype production. Longitudinal research examining the development of innovations over time (Schroeder, Van de Ven, Scudder & Polley, 1986; Van de Ven, 1986) challenges models that attempt to fit the process of innovation into discrete and sequential stages. Adapting Kanter's (1988) framework, we will analyze the politics of radical innovation from the perspective of exploring major tasks inherent in the process which are critical for its success. These tasks are (1) project conceptualization; (2) gaining sponsorship; (3) coalition building; and (4) product development.

PROJECT CONCEPTION

Radical innovations are born when an engineer recognizes the potential for developing a new technology with dramatically different performance features. However, the real challenge for the innovative engineer will be in winning legitimacy for the radical innovation. In this stage the innovative engineer(s) must secure the time and resources necessary to develop the idea to the point where it can demonstrate enough potential to attract the interest of potential sponsors (Burgelman & Sayles, 1986; Dyer & Page, 1988). Until it is "appreciated" (Van de Ven, 1986) and "raised over the threshold of consciousness" (Kanter, 1988) its future is limited. As Burgelman and Sayles

(1986) note, high tech companies like United are constantly dealing with unsolicited innovative projects from their talented engineers, which they term "autonomous strategic initiatives." The problem lies in deciding which projects should receive funding and encouragement.

Generally, radically innovative projects are not popular and are seldom solicited by management unless the company is experiencing a crisis caused by periodic environmental discontinuities such as an emergent new technology or unexpected legal, social or political events (Tushman & Romanelli, 1985). In the absence of an environmentally driven push for dramatic corporate "reorientation," radical innovations must overcome two strong countervailing forces: organizational inertia and perceptions of risk.

Organizational Inertia

Research on organizations demonstrates that with growth and maturity comes a progressively stronger resistance to radical change—organizational "inertia." Organizational members become increasingly uncomfortable and resistant to radically new technologies because they value existing internal and external relationships, and may have become overspecialized.

Internal and External Relationships

Close, comfortable relationships develop with important customers, suppliers, capital backers, government agencies, and others, and internally across different functional units. These relationships become familiar and valued, and represent a considerable investment of time and resources. Radical innovation often changes these relationships, may redefine the organization's strategic context and "mission," as well as introduce ambiguity into stable power relations. The longer these relationships persist without major change, "the more complex these social and normative outcomes become, and the more multilevel commitment processes are a source of resistance to change and inertia" (Tushman & Romanelli, 1985, p. 192; Schein, 1985).

Overspecialization

Competence becomes defined in terms of specialization in a particular technology, and these "specialties" are defended by stable, self-perpetuating groups of engineers with a vested interest in maintaining the status quo. As specialization increases, information search and processing of alternative technologies becomes restricted and increasing reliance is placed on prior knowledge, compromising the organization's ability to recognize and adapt to promising technological alternatives emerging in radical innovation (Katz & Allen, 1985; Moore & Tushman, 1982; Tushman & Romanelli, 1985).

Perceptions of Risk

Beyond tendencies towards inertia, innovation, especially radical innovation is intimidating, and managers are fearful because of the uncertainties it poses (Schon, 1981). Radical innovations are (1) unpredictable activities, making schedules and timetables uncertain, (2) novel, with "little or no experience base to use to make forecasts about results" (Kanter, 1988, p. 171); and, (3) expensive, defying cost budgeting and offering an uncertain long term payoff (Ansoff & Stewart, 1967; Burgelman & Sayles, 1986; Mansfield, Rapoport, Schnee, Wagner & Hamburger, 1971; Quinn, 1985; Schon, 1981). Further, to foster radical innovation one must deliberately loosen managerial control systems designed to contain risk, allowing for high levels of slack in resource allocation and time management, and loose, informal structures permitting high levels of autonomy (Bourgeois, 1981; Cummings, Hinton & Gobdel, 1975; Hambrick & Snow, 1977; Kanter, 1988). A business world that tends to revolve around quarterly reports can be a very threatening place for engineers and managers advocating long-term investment in radical innovations (Pearce & Page, 1988). As Quinn (1975, p.79) notes, "Time horizons for radical innovations make them essentially 'irrational' from a present value viewpoint."

Consequently, Burgelman and Sayles (1986) note that even when managers have discretionary budgets, they often choose to fund less radical forms of innovation. They quote one manager at United:

> My charter says that 15 percent of all funds should be spent on exploratory research. That sounds great, but actually it is the toughest thing in the world. It is difficult to decide how much resources to allocate to broad fields rather than to areas where there are clear business objectives (1986, p. 22).

Vicious Cycles

Radical innovation can get trapped in a vicious cycle—because the product does not meet conventional expectations concerning technological design or intended market, managers will be reluctant to fund further exploration unless the technology can convincingly demonstrate its potential. However, funding is required to develop the technology to that point, and it is difficult to show results without adequate resources (Burgelman & Sayles, 1986).

At MGI we interviewed a junior engineer caught in that trap. While he enjoyed working in a company he regarded as exciting and innovative, he nonetheless felt stifled. Repeatedly he had conceived of promising new innovations, which, despite support and enthusiasm from his work group, were inevitably denied resources. Most often they were simply ignored by higher management. He bitterly asked,

> Is the company saying, "We want good ideas at some sort of level, but not too many and not too divergent?" Is that what they want? If so, then maybe some people wouldn't want to stay here, or even to have come in the first place.

Bootlegging

Both MGI and United recognize the need for fostering radical innovation, and attempted to break this vicious cycle through traditions of "bootleg" or "nonprogrammed" research—cultural values that tell innovative engineers to simply "steal" what they need. By encouraging such appropriation, managers compensate for their limited ability to judge the real merits of proposed innovations, especially radical innovation, by funding it informally. "Such projects also allow the laboratory to treat the more reasonable and safe projects in one way (formal funding) and the more risky one is another (informal funding)" (Burgelman & Sayles, 1986, p. 27). Organizations that value and expect innovation also may set aside "innovation banks" and "internal venture capital" to provide funds for innovation outside those in the operating budget (Kanter, 1988).

Interpersonal Networks

Beyond bootlegging, there is another strategy that can be successfully employed to better the odds. Interpersonal networks and relationships can help the engineer to overcome the forces against radical innovation (Dyer & Page, 1988). Relationships of friendship and trust with managers help radical innovations become "appreciated," and convince managers to allow the innovative engineer to devote more time to the innovation.

GAINING SPONSORSHIP

Eventually innovative engineers reach a point in the development of their idea where slack or bootlegged resources and personal time are no longer enough. To progress further, the project will require more manpower, more resources, and perhaps even specialized equipment for development and testing (Burgelman & Sayles, 1986; Dyer & Page, 1988). In order to secure this, gaining sponsorship from senior level engineers and managers is critical. Sponsors both actively advocate the innovative project and shield it against destructive forces such as budget and resource limitations. Without such sponsorship, innovation, especially radical innovation, is very difficult to foster (Frost & Egri, 1988; Hage & Dewar, 1973; Kanter, 1988; Kelly, 1976; Maidique, 1980; Nayak & Ketteringham, 1986; Pinchot, 1985; Quinn, 1979, 1985).

Timing

While gaining a sponsor for a radically innovative project is always challenging, the timing of such efforts appears to be critical (Burgelman & Sayles, 1986; Dyer & Page, 1988). Clearly, the best time to seek sponsorship for such projects is during periods where the firm is facing discontinuous environmental changes, feels the need for reorientation, and is actively trying to foster radical innovation (Moore & Tushman, 1982; Tushman & Romanelli, 1985). Unfortunately, these periods tend to be too infrequent and unpredictable to give innovative engineers much confidence in waiting.

In the absence of environmental discontinuity, timing involves seeking sponsorship during "lax" periods between major projects. To solicit sponsorship during periods of intense organizational activity around established technologies is clearly folly (Dyer & Page, 1988). Sponsors neither have the time, resources nor energy to take on such a project until the pressure eases. Note these comments of a United researcher concerning the importance of timing.

> For me "dreaming" is important. You "moonlight"; you don't quit; you keep developing the story, but do not push at the wrong time. (Burgelman & Sayles, 1986, p. 26)

Dual Authority Systems

These sponsorship roles may be integrated into an informal set of dual systems for resource allocation and project development. Functionally specialized structures and formalized procedures are used for the tasks for which they are designed—they are relatively efficient and effective in handling the funding and development of product enhancements and incremental innovations. However, to foster radical innovation, a traditional authority system is also established (Pearce & Page, forthcoming).

This traditional authority system consists of the founder (or key executive) and a select group of senior engineers and executives who have been endowed with considerable slack resources to promote radical innovation. They are both deeply committed to a shared sense of corporate mission and technological vision and have been successful in developing previous radical innovations. Given that the time lines and uncertainties involved in radical innovation make it inherently incalculable (Delbecq & Weiss, 1988; Pearce & Page, forthcoming; Quinn, 1985) they are not expected to allocate resources on the basis of a careful cost benefit analysis, but on the arational "faith" that they will know it when they see it—that they will have an instinctive feel for successful and promising technologies as they did in the past (Pearce & Page, forthcoming). Quinn (1985) quotes such a manager describing this decision making process:

> Anyone who thinks he can quantify this decision is either a liar or a fool. . . . There are too many unknowables, variables. . . . Ultimately one must use intuition, a complex feeling, calibrated by experience. . . . We'd be foolish not to check everything, touch all the bases. That's what the models are for. But ultimately it's a judgment about people, commitment, probabilities. . . . You don't dare use milestones too rigidly. (p. 83)

These powerholders have the influence and resources to successfully protect radically innovative projects from the forces in formalized systems that normally undermine them. The importance of such sponsorship cannot be underestimated. For example, Maidique (1980) found that in Digital Equipment Corporation, an innovative project had become "mired in red tape" until a powerholder attended a project review and openly questioned why such a promising project could find so little support. Afterward, the project champion noted:

> Suddenly the barriers to my project came down. What normally might have taken a year or more to complete became a six-month project. (Maidique, 1980, p. 68)

Interpersonal Networks

Unfortunately, the relative isolation of executives undermines the purpose of traditional authority systems (Quinn, 1985). Their effectiveness is limited to the scope of the interpersonal networks of the powerholders. Ironically, those within such networks are seldom the creative, innovative engineers who would most benefit from such sponsorship. These executives usually deal with senior managers whose administrative duties and "technical obsolescence" have often pulled them away from the leading edge of technology (Burgelman & Sayles, 1986; Maidique, 1980), and senior engineers who have established the current technology and are heavily invested in maintaining it (Katz & Allen, 1985). Thus traditional systems can become isolated from the junior engineers who could champion radical innovation, and thus become impotent to foster it.

Ironically, some of the most innovative engineers may choose to isolate themselves from these networks as well. Brilliant and innovative engineers are often characterized as being arrogant, unreasonable and highly idiosyncratic (Burgelman & Sayles, 1986; Kenney, 1986), and tend to ignore or bypass superiors who are not on their level of competence (Dyer & Page,

1988). In contrast, the traditional authority system described above is not based on technical competence and acknowledges that these senior people may be technically obsolete, and no longer on the leading edge (Burgelman & Sayles, 1986; Maidique, 1980). These systems work on the premise that experience, intuition and personal relationships with the engineers developing new ideas are more important than technical competence in making this type of funding decision (Delbecq & Weiss, 1988; Pearce & Page, forthcoming; Quinn, 1985). This has the unfortunate effect of undermining the personal projects of engineers who may be technically brilliant but interpersonally inept. Because such innovative engineers are either unwilling or unable to develop interpersonal relationships with key managers and senior engineers, these key players are often unwilling to sponsor their projects (Dyer & Page, 1988).

In summary, we agree with Frost and Egri (1988), who note that administrative innovation, such as dual authority systems, is required to support and sustain radically innovative projects. Further, this kind of managerial innovation and the flexibility it introduces (Quinn, 1985) often depends upon multilevel interpersonal networks within the firm.

BUILDING COALITIONS

After gaining sponsorship, attention turns to project development and analysis in preparation for an organizational "rite of passage"—the formal project review. The project review board typically consists of representatives from a variety of major departments. Given the diversity of this board, building a coalition by selling the project to potential allies is critical. "Social and political factors, such as the quality of the coalition building, may account for as much or more than technical factors, such as the quality of the idea, in determining the fate of innovation" (Kanter, 1988, p. 185; see also Frost & Egri, 1988). Coalition building typically involves careful lobbying of key members of powerful departments and constituencies, and the solicitation of feedback concerning what technical features of the project are poorly received, which are positively regarded, and what other features might be included to better its chances for approval (Dyer & Page, 1988).

> Cautious but astute organizational champions make sure that the causes [i.e., new ventures] they sponsor are consistent with the existing predispositions of top management. More brilliant or perhaps more risk-prone executives seek to change the disposition of top management and get them to accept a new business field as legitimate for corporate development. (Burgelman & Sayles, 1986, p. 143)

Incremental innovations, which produce variation in product technologies, usually fit within established organizational strategic, political and cognitive domains. Accordingly, the political and social processes involved in building coalitions are "not very complex" because the technology is familiar (Normann, 1971).

Radical innovations, on the other hand, cause the organization to "reorientate" around them—perceptions of the environment, the definition of the organization's strategic domain or context, political and power relationships between subsystems and coalitions—all may be dramatically affected (Kanter, 1988; Normann, 1971; Tushman & Romanelli, 1985). Thus the project review process becomes increasingly political as the projects under consideration become increasingly radical. The more radical the proposed innovation, the more it threatens to change established political processes and power relations between groups. Normann (1971, p. 205) cites the "Drugs and Pills" corporation as an example of the impact new innovative research efforts can have on the values, power relations and patterns of interdependency between different organizational groups and coalitions over time:

> Research at Drugs and Pills had been conducted only in the field of chemistry and had been carried out only by chemists and technical engineers. Then two pharmacologists were employed, who eventually persuaded the managing director to direct the research efforts of the company to pharmacology. Two such projects were undertaken, and a few years later the whole structure of the company had changed: the new pharmacological laboratory was the largest building, a large share of the people employed were pharmacologists, and the two persons who had initiated the change had become marketing director and research director.

Technical Ideologies

Beyond their political impact on intergroup relationships, technologies also are an integral part of the underlying value structures of different groups. Research and development, product engineering, marketing and manufacturing, by virtue of the different functional specializations they pursue, tend to develop different goals, values, norms and even language (Alderfer & Smith, 1982; Lorsch & Lawrence, 1965; Schein, 1985). Consequently, these departments often develop quite different definitions concerning the nature of innovation, the thrust of R&D, the role of quality engineering, and the types of products the company should be pursuing. In building coalitions, it is critical to understand their perspectives and positions on the type of technology being advanced, as well as the recent political history between the departments, who frequently hold each other in contempt and periodically try to dominate each other. We will illustrate these perspectives by comparing the perspectives of three divisions at MGI: the CAD, CAT and Manufacturing divisions.

CAD Division

The Computer Aided Design (CAD) division is famous for R&D and innovation. Its research engineers have some of the most demanding definitions of innovation at MGI. For them, real innovation is radical innovation—a dramatic technological breakthrough with stunning performance capabilities. They do not recognize more incremental forms of innovation, such as product enhancement, as being innovative at all. At one extreme, one senior research engineer defines innovation solely in terms of quantum technical leaps, such as going from slide rules to calculators (Dyer and Page, 1988).

For CAD engineers, the role of engineering is to identify and pursue these leading edge breakthroughs, revolutionizing and creating new markets. One of the founders, John Parks, is a legend among the research engineers.

> John goes to the market and identifies needs that the market isn't even aware of. Then he goes back to the lab and designs something that fills that need. It's a case of using a meta-level, being aware of needs and transcending that need. (Dyer & Page, 1988, p. 28)

In short, for CAD engineers quality engineering is advanced technology, regardless of market success.

CAT Division

The Computer Aided Training (CAT) division, in contrast, defines quality innovation as enhancement and incremental improvement of existing products and technologies. From inception, this division has been driven by market pressures and customer needs. The division's president describes his philosophy:

> I've never sat down and invented something. It's always been somebody coming in with a problem they want solved. There are times when we drove the technology. But the idea for better products was pushed by the market.(Dyer & Page, 1988, p. 29).

For the CAT division, quality engineering is market-driven, close to the customer, and at least two steps ahead of the nearest competitor. For these engineers, market success is the ultimate measure of quality engineering.

Manufacturing

This department tends to define quality innovation in terms of process innovation, which increases the reliability or lowers the cost of a technology without necessarily affecting its performance capabilities. They note that it is one thing to design an advanced technology, and quite another to build it in large numbers.

For manufacturing engineers, quality and reliability are admired above technical sophistication. They note that MGI's major customers demand high-performance and high quality systems in return for the premium prices they pay. Consequently, they see themselves in a critical "guardian" role, insuring that technologies are workable and reliable before they are shipped.

In summary, we agree with Normann (1971), who explores innovation as a political process by viewing organizations as systems composed of competing subsystems and coalitions, with the innovative product as a political artifact—a tangible result and physical manifestation of the interaction of different subsystems. "Thus, its set of dimensions and their relative importance may be considered a mapping of the value and power structures of the organization, and the new set of product dimensions must relate to the existing or changed values and power structure of the organization" (Normann, 1971, p. 205).

Case Study: The Eclipse Project

Eclipse is an ideal project to explore the politics of developing a radically innovative idea from conceptualization through project review, for the technologies and markets involved directly contradicted most of the conventional technical, strategic and political traditions at MGI. Eclipse is the saga of a group of determined engineers trying to break into the work station market. Apollo, Sun, Digital Equipment and other firms offer CAD/CAM work stations, and some people at MGI felt that they could deliver a product with much better graphics capabilities. However, the political climate was against such a move. Successfully breaking into an established, competitive work station market was not assured, and represented a strategic risk that did not appeal to CAT division managers. Beyond strategy, the radically new technologies involved were "peripheral" to more mainstream technical efforts, and therefore not "sexy" enough to be of much interest to many senior CAD division engineers. Further, such efforts defied precedent. The first attempt design a work station failed—it was never formally approved, the product was judged as technically deficient, and the senior manager responsible was fired. The president clearly indicated that MGI was not in the work station business.

However, some of the engineers involved refused to give up. By coming in early, and bootlegging time from other projects, they eventually developed the idea for a low cost work station with dramatically enhanced graphics capabilities. This project was called Pandora, after Pandora's Box, in honor of the discord it was bound to loose in the organization. Their project lived up to its name, generating such opposition that they were "thrown out on their ears" at their first project review. Senior engineers from both divisions thought it was just "poor technology," strongly objected to the idea of a work station, and criticized it as being "not different enough from the competition." In short, it was judged "not worthy of MGI."

In retrospect, the engineers are convinced that they had assembled enough technical information at the time of that project review for its approval. Given the eventual enthusiastic approval of the technology, these harsh reviews were not an objective reaction to the technical merits of the proposal; they reflected political views and prejudices concerning the types of markets, technology and innovation involved. The engineers had not prepared the political support they would need to buck conventional technical politics. They realized that their technology could not be "objectively" reviewed until these political issues had been taken care of. Without such political strategies and the support of key powerholders, their project, regardless of its technical merits, was doomed to fail.

First and foremost, the engineers had not recruited CAD division support for the unconventional, controversial features of their radically new technology, especially for their process innovations, which offered cost advantages without increasing performance. MGI's products tend to be high tech, high performance and high cost. Further, they used a new "X" type display in a division that holds such displays in disdain, breaking a long standing tradition of "Z" type displays in CAD products. Without the support of key senior CAD engineers, top management demanded an exhaustive defense of these features—a defense they did not have time to prepare. They had not researched the limitations of other technical approaches, and thus could not defend their position from alternative approaches that used technologies with which MGI was more familiar and comfortable.

Further, they were proposing a product in a "banned" market. The president is the key decision maker at MGI, and they had no key senior CAT division officials on their side to convince him of the merits of the work station. The president clearly did not like work stations. Further, the CAT division demanded market projections and a business plan, without which they refused to proceed. Moreover, the engineers had done no lobbying and had no sponsorship.

However, the proposal was not dead. Top management had recently reorganized the company, and had given strategic priority to new product development. At this point, two senior CAD engineers saw some potential in the project, and began to sponsor this small group, giving them the time, and engineers "on loan" to continue research. Technically, the team knew they had the right approach, and after further research and reviews of other architectures, they began to understand why. Further, the cost of high quality "X" type displays that could produce "Z" quality lines was dropping consistently. Politically, the engineers, under the direction of their sponsors, began an intense lobbying effort with the president and top management of both divisions to generate support and to solicit feedback. They repackaged the project as a graphics terminal add on, removing the work station stigma, although the product could easily be expanded into a full work station at some later, more politically expedient date. They renamed the product Eclipse, and incorporated technical features that were "pet" technical interests of the president. They also prepared lots of pictures of their "X" system's line drawing capability, hoping to visually overcome the traditional bias against "X" type systems. Before the next project review, their CAD sponsors led the engineers through three days of mock reviews, trial arguments, practice presentations and grueling technical questions. While they were presenting virtually the same technical information as before, this time they felt politically ready.

The president loved the pictures and favored the technology (which had been modified to his specifications). Engineers could answer all the technical questions this time around and their target market was no longer objectionable. Further, many senior engineers from both divisions had been successfully lobbied and were favorably impressed, creating a powerful positive coalition for the project, and they had formal sponsorship from the two senior engineers in their area—all of the political contingencies had been met. The president, enthusiastic at the prospect of finally seeing some of his technical interests developed, felt this project was not only technically sound but politically expedient—it would encourage new product development by demonstrating that personal projects of innovative engineers could become products. The project was approved.

PROJECT DEVELOPMENT

In project development, the politics involved in coalition building goes from the general level of technical design in the project review, to the specific, operational level of actual product development. This involves coalition building at many levels and in many departments. Burgelman and Sayles (1986, p. 74) note:

> The product champions have the challenges and frustrations that are characteristic of any project manager's job. They must deal with a variety of groups over whom they have no control, each of which may be critical to the project's success and each of which has different and often contradictory goals.

Boundary Spanning

One of the hardest transitions in the product development process involves the innovative engineer who championed the product, and now has assumed the formal role of project or product manager. The types of skills necessary for being a successful product champion in the initial stages of innovation are quite different from the skills necessary for successful product development.

Effective product champions are characterized as fiercely autonomous, stubborn, tenacious, arrogant and insensitive to the demands of other groups and projects (Burgelman & Sayles, 1986; Dyer & Page, 1988). These are the types of attitudes and skills that are critical initially—without them needed time and resources would not be "appropriated," and the daunting task of introducing a radically new idea into an unsupportive environment would probably have not been undertaken in the first place. This point is epitomized by the senior engineers and managers of Data General's radically innovative Eagle computer group, that explicitly used "cockiness" as a central selection criteria in recruiting the best and the brightest to work on their project (Kidder, 1981, p. 65).

This trait is further reinforced by the fact that most established groups and coalitions tend to oppose radical innovation. Because R&D engineers are typically well-educated in their specialities, they can advance and defend their opinions with a variety of arguments and "facts," while undermining, if not ridiculing alternative positions. Consequently, this adds to their tendency to come across as unreasonable, patronizing, egocentric, stubborn and nonresponsive (Burgelman & Sayles, 1986; Kenney, 1986). In his case on biotechnology, Kania (1982, p.17) notes these comments of one corporate analyst:

> Academia places a clear premium on irrationality and antisocial behavior. Many of our best scientists have spent little time developing their interpersonal skills and can be very difficult to control.

However, the role of product or project manager is explicitly boundary spanning. The success of the project depends on the successful transfer of a complex and ambiguous radical new technology through a variety of interdependent groups. Innovative engineers must go from product champions to organizational diplomats. They must find ways to communicate with groups they formerly ignored, build relationships of openness and trust with people they formerly held in contempt, and recognize and accommodate the concerns and interests of groups they formerly defied. While their sponsor initiated this process in building coalitions for project review, the burden of frequent interaction and interdependence now falls on their shoulders. Not surprisingly, this "history" between groups tends to result in conflict more than collaboration:

> What is crucial is the obvious requirement that professionals from each of the . . . groups must work more closely than these divergent tendencies would normally encourage.

> . . . Each group must be willing to try out and explore modifications that would suit the constraints the other operates under or believes to exist. Such patterns of mutual accommodation and the trust to admit or concede that a favorite parameter might be dropped or may be a variable occur only between employees who share common group membership and common loyalties. . . . the inherent frustrations of this amorphous period when so much is up for grabs can only encourage each group to scapegoat the other or to insist that the other side must make all the concessions or is "bull-headed." (Burgelman & Sayles, 1986, p. 69)

Needless to say, this task is difficult, but critical (Lorsch & Lawrence, 1965). MGI's history is littered with the wreckage of the projects of champions who were not able to make a successful transition to product manager, or find someone to fill that role.

Unfortunately, formal efforts to facilitate collaboration typically fail with radically innovative projects. Cross-departmental representation in all phases of the project is not feasible, since initially that type of input is exactly what the radical technology needs to be buffered and protected against. Matrix structures and other formal coordination mechanisms also do not tend to work well, being unable to handle the magnitude of ambiguity and conflict characteristic of radical product innovation. For example, to facilitate collaboration on a radically innovative project at MGI, management instituted a matrix structure arrangement and frequent intergroup meetings. The result was, in the words of one senior engineer, "a terrible mess—an absolute morass." He further commented:

> What would happen is that we'd get together in meetings, sort of agree on some compromises, then go back to normal after we left the meeting. The manufacturing people would go back to their rules; the engineering people would go back to changing their drawings every three hours; the product management people would go back to missing possible design opportunities in order to keep the customer happy. It was a cesspool of noncommunication.

Interpersonal Networks

Researchers consistently note that interpersonal networks—informal communication channels usually based on preexisting friendships with members of the other groups—prove critical in successful product development (Burgelman & Sayles, 1986; Delbecq & Weiss, 1988; Dyer & Page, 1988). Many MGI project managers admit that without interdepartmental friendships, many innovative projects, especially those involving radically new technologies, would never be built. For example, one engineer described how he resolved a major problem of excessive documentation requirements from the manufacturing department:

> I asked the manufacturing manager why it was so difficult to get things through. He explained that he had to be strict. They were aware of the customer's quality standards. One malfunction or missing spec, and he knew he'd be fired. So I asked him if he'd feel good about having a group of our engineers sign a statement saying something like: "To the best of our knowledge, this equipment is free from flaws and will perform properly." He said, "O.K." It took him off the hook. From then on things went a lot smoother. If it weren't for my friendship with the manufacturing manager we might never have delivered that system. He really stuck his neck out for me. (Dyer & Page, 1988, p. 39)

Cooptation

The product development stage is the first time other groups and relatively large numbers of "outsiders" have been involved with the design and development of the technology. This creates the opportunity for other groups or individuals to "coopt" the radical technology, causing it to lose much of its technologic originality and become an idiosyncratic innovation or a mere enhancement of an existing technology.

Idiosyncratic Cooptation

When the project enters the development stage, in addition to being assigned a budget, it is assigned several senior engineers to assist with technical problems and oversee the project's progress. Consequently, the project manager is forced to work with and often accept the ideas and advice of the senior engineers, which can result in delays and rework (Dyer & Page, 1988). In extreme cases the project becomes coopted by powerful senior engineers and becomes little more than a forum for their own idiosyncratic agendas. In less extreme cases development is channeled by these senior members, and certain "pet interests" become incorporated as additional features. For example:

> The company developed an interest in a high-technology area in the biological field. The efforts were directed by a Ph.D who some years before had developed an innovative method of separating materials at the molecular level. Such separation was an important step in producing certain new products in which we became interested. But, as we later realized, Dr. X's technique, as interesting as it was, wasn't the best way of making the transition to large-scale manufacture. But as long as Dr. X was in charge, alternative methods would not be explored. (Burgelman & Sayles, 1986, p. 38)

Systematic Cooptation

Historically, MGI engineers with skills crucial to project conception and innovation have not been well-suited to the "rigors" of development. They like to play with new ideas, and do basic conceptualization and system architecture, but they prefer to have someone else do the detail work involved in product development (Dyer & Page, 1988). However this lack of interest allows individuals and groups who may not completely understand or support the "vision" of this radically new technology to slowly, but surely change its character through thousands of small development decisions. In this way the design is incrementally moved towards familiar designs and conventional solutions for these types of technical problems. This occurred in at least two major projects at MGI. In both cases, the innovative engineers, unwilling to become more deeply involved in the detail work and thus unable to stem the tide of incremental design decisions, left the company in disgust halfway through the project. In short, maintaining the character of radical innovation requires ongoing involvement of the initial group, continuity in group membership, and a means of communicating, if not socializing, new participants to the vision of the new technology to hopefully produce commitment to that design (Kanter, 1988).

Buffering

Another critical task is buffering the project development team from dysfunctional levels of pressure and scrutiny from customers and top management (Burgelman & Sayles, 1986; Dyer & Page, 1988; Kidder, 1981). This is critical to insure that resources continue to flow to the group, and that group efforts are not disrupted. While buffering can get out of control, and lead project managers to deceive top management and important customers concerning the progress of a project (Burgelman & Sayles, 1986; Dyer & Page, 1988), when managed well it can be very beneficial.

> X had to protect his project from too close scrutiny from management. He had to do a lot of tricks to get the time necessary to really develop it. It was a struggle for many years. But he persisted and ended up with one of the best-protected processes in the field. (Burgelman and Sayles, 1986: 80)

IMPLICATIONS

Despite the preliminary nature of this analysis, two implications present themselves so consistently they merit special discussion and consideration. First, we will discuss how different political processes affect radical innovation, arguing certain forms of political activity can be a powerful force facilitating innovation. Second, we will explore the role of interpersonal networks in the positive political processes which facilitate radical innovation.

The Legitimacy of Political Processes

The role of politics in the process of innovation is typically denigrated, if not attacked by the engineers at MGI. Likewise, Frost (1988) notes that researchers often define organizational politics as "illegitimate activity" (House, 1984; Mintzberg, 1983). Concerning innovation, this bias seems justifiable, for politics is frequently associated with "red tape" and the capricious use of power to support politically expedient but technically unworthy technologies and "sandbag" [ignore] exciting, radically innovative efforts. Political processes are never mentioned in a positive sense—they are perceived as detracting, not contributing, to the process of innovation.

This is ironic, given that our research suggests that many of the difficulties associated with politics develop because radical innovation falls outside of established political processes. Within the bounds of normal politics, development efforts take advantage of preexisting sets of norms, standard operating procedures and communication channels for effective negotiation and collaboration between different individuals and groups. Falling outside these bounds, radical innovation cannot tap into these informal, political channels designed to increase the effectiveness of intergroup interaction and collaboration. These informal, political processes are based on carefully negotiated sets of understandings and expectations that are often violated by the degree of uncertainty, ambiguity and change inherent in radical product innovation. Consequently, groups are far more likely to demand that formalized bureaucratic procedures and extensive documentation govern intergroup interactions when a radical innovation is introduced. This degree of formalization acts both as a defensive strategy to minimize group risk (given a significant probability of project failure) and as an offensive strategy to undermine the project (being a very slow, ponderous, resource intensive and time-consuming way of interacting). Consequently, radical product innovation tends to suffer from the lack of necessary political processes.

Further, research has begun to identify political processes established to foster radical product innovation. Sponsorship roles and dual authority systems are examples of political activities that nurture, buffer and protect vulnerable projects until they are prepared for the project review. Admittedly, these types of roles and authority systems are not well understood. They have only recently evolved from "informal" status to recognition and legitimacy, as the importance of fostering continuous innovation became apparent for American industry. However, their importance can not be minimized, since these political activities appear critical to success. Further research on the nature of these positive political processes is clearly needed.

Table 1. Interpersonal Relationships and Radical Innovation

Task	*Critical Interpersonal Relationship*	*Function*
Conceptualization	engineer-manager	Gain more time and resources for initial concept development
Gaining Sponsorship	engineer-sponsor	Gain resources, staff, equipment for intensive project development; Gain access to traditional authority system
Project Review	sponsor-top management	Coalition building
Product Development	product manager-departmental contacts	Effective intergroup cooperation and collaboration

Interpersonal Relationships

One critical component in developing such positive political processes is viable, multilevel interpersonal contacts between organizational members. These contacts develop into interpersonal networks between project managers and key actors in different groups. Given a favorable organizational "atmosphere" or culture, individuals can develop personal relationships of mutual respect and trust across organizational groups. These interpersonal networks provide a foundation of trust, open channels of communication and an expectation of mutual benefit that allow different groups to quickly develop understandings around unique work arrangements and novel technologies. These understandings, while precarious, effectively substitute for political arrangements and do permit resources and collaborative efforts to be utilized more effectively. Project managers at MGI invariably cite preexisting relationships with members of other groups as being the deciding factor between project success and failure. In most circumstances, only these interpersonal bonds allow other groups to overcome their initial negative reactions concerning the risk presented by the project and the vulnerability effective collaboration creates. This is clear in the preceding analysis of each of the critical tasks facing radical innovation, from conceptualization through prototype development, where interpersonal relationships prove to be of central importance. This information is summarized in Table 1.

Recent research findings by Eisenhardt and Bourgeois (1988) on the politics of strategic decision making in rapidly changing, "high velocity" environments suggests that, in certain contexts, interpersonal networks can actually replace political activities. Their study of Silicon Valley microcomputer firms revealed that successful firms featured a supportive decision environment characterized by open communication, trust and mutual respect, despite the ambiguities and uncertainties posed by both the proposed technologies and the industry environment itself. In contrast, firms engaging in political processes were generally unsuccessful—the restricted information flow, time investment in lobbying activities, and centralized power structures characteristic of the more political firms proved too slow and debilitating in highly dynamic environments. The premise that political behaviors can ultimately be transcended given the right culture, leadership and positive interpersonal relationships is an intriguing one, and merits further study.

CONCLUSION

This chapter has attempted to explore the inherently political nature of radical innovation. Given the dearth of published findings in this area, more extensive, systematic, empirical research is clearly needed. While the research analysis advanced in this chapter is tentative and exploratory, it will hopefully serve to stimulate further discussion and research into the political dimensions of innovative processes.

REFERENCES

Abernathy, W.J., & Clark, K.B. (1985). Innovation: Mapping the winds of creative destruction. *Research Policy, 14,* 3-22.

Abernathy, W.J., & Utterback, J.M. (1977). Patterns of industrial innovation. *Technology Review, 80,* 58-64.

Alderfer, C.P., & Smith, K.K. (1982). Studying intergroup relations embedded in organizations. *Administrative Science Quarterly, 27,* 35-65.

Ansoff, H.I., & Stewart, J.M. (1967). Strategies for a technology-based business. *Harvard Business Review, 45,* 71-83.

Bourgeois, L.J. (1981). On the measurement of organizational slack. *Academy of Management Review, 6*(1), 29-39.

Burgelman, R.A., & Sayles, L.R. (1986). *Inside corporate innovation.* New York: Free Press.

Cummings, L.L., Hinton, B., & Gobdel, B. (1975). Creative behavior as a function of task environment: Impact of objectives, procedures, and controls. *Academy of Management Journal, 18*(3), 489-499.

Delbecq, A.L., & Weiss, J. (1988). The business culture of Silicon Valley: Is it a model for the future? In J. Hage (Ed.), *Futures of organizations* (pp. 124-141). Lexington, MA: Lexington Books.

Dunford, R. (1987). The suppression of technology as a strategy for controlling resource dependence. *Administrative Science Quarterly, 32,* 512-525.

Dyer, W.G., & Page, R.A. Jr. (1988). The politics of innovation. *Knowledge in Society, 1*(3), 24-41.

Eisenhardt, K.M., & Bourgeois, L.J., III (1989). The politics of strategic decision making in high velocity environments: Towards a midrange theory. *Academy of Management Journal, 31*(4), 737-770.

Fast, N.D. (1979). The future of industrial new venture departments. *Industrial Marketing Management, 8,* 264-273.

Frost, P.J. (1988). Power, politics and influence. In F.M. Jablin, L.L. Putnam, K.H. Roberts, & L.W. Porter (Eds.), *Handbook of organizational communication* (pp. 503-548). Beverly Hills, CA: Sage.

Frost, P.J., & Egri, C.P. (1988). *Is it better to ask for forgiveness than to seek permission? The influence of current and past political action on innovation in organizations.* Paper presented at the annual meeting of the Academy of Management, Anaheim, CA.

Gandz, J., & Murray, V. (1980). The experience of workplace politics. *Academy of Management Journal, 23*(2), 237-251.

Hage, J., & Dewar, R. (1973). Elite values versus organizational structure in predicting innovation. *Administrative Science Quarterly, 18,* 279-290.

Hambrick, D.C., & Snow, C.C. (1977). A contextual model of strategic decision making in organizations. In R.L. Taylor, J.J. O'Connell, R.A. Zawacki, & D.D. Warrick (Eds.), *Academy of Management Proceedings* (pp. 109-112). Mississippi State, MS: Academy of Management.

House, R.J. (1984). *Power in organizations: A social psychological perspective* (Working paper). University of Toronto.

Kania, E.M. Jr. (1982). *Case study of Advanced Genetics Sciences, Inc. Cambridge, MA: Harvard Business School Press.*

Kanter, R.M. (1983). The change masters. New York: Simon & Schuster.

________(1988). When a thousand flowers bloom: Structural, collective, and social conditions for innovation in organization. *Research in Organizational Behavior, 10,* 169-211.

Katz, R. & Allen, T.J. (1985). Organizational issues in the introduction of new technologies. In P. Kleindorfer (Ed.), *The management of productivity, technology and organizational innovation* (pp. 103-125). New York: Plenum Press.

Kelley, G. (1976). Seducing the elites: The politics of decision making and innovation in organizational networks. *Academy of Management Review, 1*(3), 66-74.

Kenney, M. (1986). *Biotechnology: The university-industrial complex.* New Haven, CT: Yale University Press.

Kidder, T. (1981). *The soul of a new machine.* New York: Avon Books.

Lorsch, J.W., & Lawrence, P.R. (1965). Organizing for product innovation. *Harvard Business Review 43*(1), 109-122.

Maidique, M.A. (1980). Entrepreneurs, champions, and technological innovation. *Sloan Management Review, 21*(2), 59-76.

Mansfield, E., Rapoport, J., Schnee, J., Wagner, S., & Hamburger, M. (1971). Research and innovation in the modern corporation: Conclusions. In E. Mansfield, J. Rapoport, J. Schnee, S. Wagner, & M. Hamburger (Eds), *Research and innovation in the modern corporation* (pp. 206-222). New York: Norton.

Mayes, B.T., & Allen, R.W. (1977). Toward a definition of organizational politics. *Academy of Management Review, 2*(4), 672-678.

Miles, R.E., & Snow, C.C. (1978). *Organization strategy, structure and process.* New York: McGraw-Hill.

Miller, D., & Friesen, P.H. (1982). Innovation in conservative and entrepreneurial firms: Two models of strategic momentum. *Strategic Management Journal, 3,* 1-25.

Mintzberg, H. (1983). *Power in and around organizations.* Englewood Cliffs, NJ: Prentice-Hall.

Moore, W.L., & Tushman, M.L. (1982). Managing innovation over the product life cycle. In M.L. Tushman & W.L. Moore (Eds.), *Readings in the management of innovation* (pp. 131-150). Boston: Pitman.

Nayak, P.R., & Ketteringham, J.M. (1986). *Break-throughs.* New York: Rawson Associates.

Normann, R. (1971). Organizational innovativeness: Product variation and reorientation. *Administrative Science Quarterly, 16,* 203-215.

Pearce, J.L., & Page, R.A. Jr. (forthcoming). Palace politics and radical innovation. *Journal of High Technology Management.*

Perry, L.T., & Page, R.A., Jr. (1986). *Why radical product innovation does not occur in large organizations: The intolerance factor.* Paper presented at the annual meeting of the Academy of Management, Chicago, IL.

Pinchot, J., III (1985). *Intrapreneuring.* New York: Harper & Row.

Quinn, J.B. (1979). Technological innovation, entrepreneurship and strategy. *Sloan Management Review, 20,* 19-30.

________(1985). Managing innovation: Controlled chaos. *Harvard Business Review, 63*(3), 73-84.

Sapolski, H.M. (1972). *The polaris system development.* Cambridge, MA: Harvard University Press.

Schein, E.H. (1985). *Organizational culture and leadership.* San Francisco, CA: Jossey-Bass.

Schon, D.A. (1981). The fear of innovation. In R.R. Rothberg (Ed.), *Corporate strategy and product innovation* (pp. 36-45. New York: Free Press.

Schroeder, R., Van de Ven, A., Scudder, G., & Polley, D. (1986). *Observations leading to a process model of innovation* (Working paper No. 48). Strategic Management Research Center, University of Minnesota.

Tushman, M.L., & Romanelli, E. (1985). Organizational evolution: A metamorphosis model of convergence and reorientation. *Research in Organizational Behavior, 7,* 171-222.

Van de Ven, A. (1986). Central problems in the management of innovation. *Management Science, 32,* 590-607.

PROPENSITY FOR CHANGE:

A PREREQUISITE FOR GROWTH IN HIGH TECHNOLOGY FIRMS

Donald L. Sexton

The high technology industry is viewed by many as today's newest high growth industry. People recognize the promise that high technology offers for future economic progress and well being (Miller & Cote, 1985). Today, the words "high technology" are once again being viewed as analogous to high growth. Such has not always been the situation. According to Maidique and Hayes (1984), during 1969-1984, the world's perception of the competence of U.S. companies has moved from a marked superiority over other nations in the ability to manage technology to one of marginal, if any, management superiority. They also defined six themes of success: business focus; adaptability; organizational cohesion; entrepreneurial culture; sense of integrity; and "hands-on" top management. The terms entrepreneurial culture, entrepreneurial instinct or entrepreneurial spirit all appear to provide a somewhat nebulous definition of an attitude towards growth and change in the organization. Others, searching for a more detailed definition, describe the high technology firm in terms of the complexity of its products and/or processes, and the intensity of change with respect to the technology used in producing the product. They further suggest that these firms undertake a process of continuous organizational redesign (Soukup & Cornell, 1987). The mystique that is called entrepreneurship is really only a psychological propensity for growth on the part of the owner, founder or CEO, and the other factors such as business focus, adaptability, organizational cohesion, organizational redesign, sense of integrity and "hands-on" top management are measures of the owner, founder or CEO's ability to manage growth or his/her approach to managing growth. In essence, the management of change or the management of growth has two major components: (1) the CEO's propensity for growth and (2) his/her ability to manage growth.

METAMORPHIC STAGE GROWTH MODELS

A number of authors feel that growth in a firm can be expected to follow a metamorphic stage growth process that evolves from initiation to maturity and/or decline. Metamorphic stage growth models have been described in growth literature by a number of authors, most of whom disagree only on the number of stages through which a firm passes (Churchill & Lewis, 1983; Dale, 1952; Davis, 1951; Drucker, 1954; Scott & Richard, 1987; Stanworth & Curran, 1976; Steinmetz, 1969), or on the changing characteristics of the organization over time (Adizes, 1979; Downs, 1967; Grenier, 1972; Kimberly & Miles, 1980; Quinn & Cameron, 1983). A typical metamorphic stage growth model is presented in Figure 1.

Patterned after the product life cycle concept, metamorphic growth stage models depict the firm as passing through a series of stages as they grow from venture initiation to maturity. The models argue that organizations are born, experience growth and then either decline or redevelop. Entrepreneurial activity is depicted as dominating the first stage. As the organization grows and matures during the second stage, the orientation shifts increasingly toward acquiring market share and defending market niche. Finally, in the third or last stage, when the product or firm maturity has passed, the firm is expected to experience decline unless it undergoes major strategic transformation or redevelopment. Hofer and Schendel (1978, p. 34) described emerging industries as roughly analogous to the introductory and early growth stages of the product life cycle and the transition to maturity as being roughly analogous to shakeout or the later stages of growth immediately preceding the maturity stage.

Many researchers have expressed disagreement as to the applicability or appropriateness of metamorphic stage growth models to emerging businesses and especially to high growth firms in the high technology industry. The appropriateness or applicability of metamorphic stage models has been questioned on the basis of: (a) the assumption that natural and social phenomena belong to the same category of entities for the purpose of developing theories and/or explanations; (b) the use of business size as one dimension and company maturity or the stage of growth as the second

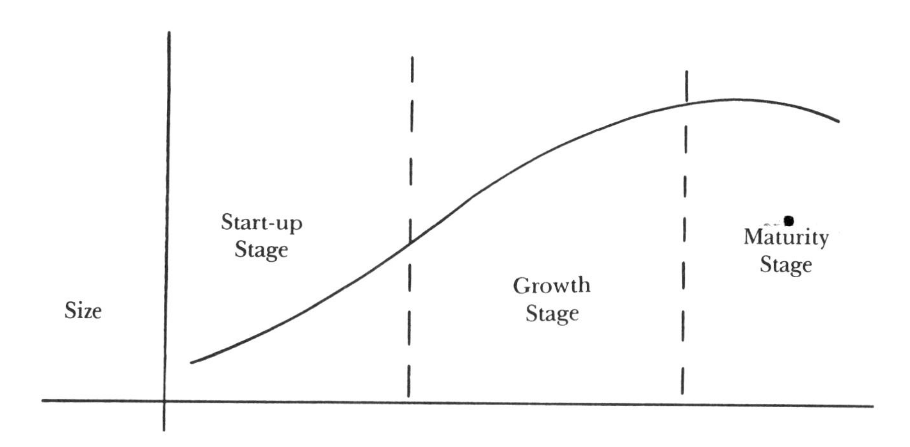

Figure 1. Typical Metamorphic Stage Growth Model

dimension; (c) the underlying use of the product life concept as an independent variable rather than considering those dependent relationships that may impact on the stage of the life cycle curve. In addition, the models have failed to accurately describe the growth patterns of new or emerging firms.

Stanworth and Curran (1976) believed that stage models were more readily applicable to larger firms. They disagreed with the assumption that natural and social phenomena belong to the same category of entities for the purpose of developing theories or explanations. In their opinion, social phenomena understand their own behavior and can act purposefully while natural phenomena cannot.

The life cycle models reflect a typical pattern of organizational development based on increasing age and size. However, there is considerable evidence for questioning this view (Churchill & Lewis, 1983; Gray & Ariss, 1985; Quinn & Cameron, 1983).

The social phenomenon point made by Stanworth and Curran (1976) indicates that the growth of the firm is under the control of the owner, founder or Chief Executive Officer of the firm. Those in control of the firm may initiate, foster, nurture or prune growth according to their own propensity or desire for growth and in accordance with their capabilities to manage it.

Churchill and Lewis (1983) were also critical of the appropriateness of metamorphic growth models.

> While useful in many respects, these frameworks are inappropriate for small business on at least three counts. First, they assume that a company must grow and pass through

> all stages of development or die in the attempt. Second, the models fail to capture the important early stages in a company's origin and growth. Third, these frameworks characterize company size largely in terms of annual sales and ignore other factors such as value added, number of locations, complexity of product line, and rate of change in product or production technology. (p. 31)

Following the criticism of Churchill and Lewis, Gray and Ariss (1985) suggested that size may be the result, rather than the determinant of certain strategic changes, and that changes in the organization's internal and external environment create the impetus for passage into a new phase or stage.

Metamorphic growth models are based on the product life cycle concept. Gardner, in describing the product life cycle concept, suggested that while the "concept reflects what many observe to be reality," a good descriptive sense of the concept has not been developed. In addition, not much is known about treatment of the concept as a dependent or independent variable, and knowledge of the concept concerning the earlier stages of the life cycle seems to be most limited (Gardner, 1986a, 1986b). Gardner further suggested that what is needed is a focus on the product life cycle concept as a dependent variable and an examination of those dependent relationships that appear to have potential for determining the shape of the life cycle curve (Gardner, 1986b).

Organizational life cycle theory may be questioned on the basis of suspected underlying concepts. Perhaps the theory cannot be accurately applied. Some firms experience such rapid growth that they appear to pass over some defined stages on their way to becoming large organizations. Others achieve success and then seem to disappear. Some continue to grow, others plateau and some experience only limited growth. In addition, there seems to be no assurance that success in one stage assures movement into the next stage nor is it even a prerequisite. Further, it appears that maturity or decline, as a final stage, may be prolonged or avoided through product innovation, geographic expansion or other initiatives.

Given the limitations of the metamorphosis models, it can be concluded that their appropriateness or applicability may be limited because: growth is a social and hence a controllable variable; their dimensions are inappropriate; their applicability to the earlier stages of a business venture is limited; and they use the product life cycle as an independent rather than a dependent variable.

GROWTH: NOT A COMMON PHENOMENON

The applicability of metamorphosis models to new or emerging firms can also be questioned on the basis that significant growth occurs in only a limited number of firms. In 1986, the 17 million plus small businesses represented roughly 99 percent of the total number of businesses in the United States. Yet, fewer than 40 percent of these firms had total annual revenues of $25,000.00 or more, less than 10 percent had $100,000 annual sales range, and less than four percent reached the 1 million dollar plateau (*The State of Small Business,* 1987). Further, less than one percent of all nonfinancial U.S. corporations ever achieve a sales volume of $25 million per year (Cavanagh & Clifford, 1984). In addition, fewer than 650,000 firms in the United States qualify as growing firms and the top 15 percent of these firms (97,500) account for 98 percent of all gross new jobs created (Birch, 1987).

High technology firms have enjoyed a special position with regard to the media attention given to entrepreneurial firms in general. In fact, they are seen as offering more opportunities with respect to growth, regional economic development, technological innovation, higher success rates and vitality and flexibility in the economy (Cooper, 1986). In addition, 183 firms on the 1987 Inc. magazine's list of America's Fastest Growing Private Companies could be classified as high technology related firms. Firms on this list averaged a 13-fold increase in sales ($960,000 to 12.7 million) and a five-fold increase in number of employees (19 to 106) during the period 1982 to 1987 ("The *Inc* 500," 1987). Further, the Arthur Young High Technology Group found similar growth rates in their study of 213 biotechnology companies (*Arthur Young High Technology Group,* 1987).

It is also apparent that growth does not occur as a natural phenomenon. Not all firms grow. In fact, most experience only limited or marginal growth beyond their first year sales (Reynolds & Freeman, 1987; Reynolds & Miller, 1988). Growth is a social phenomenon that occurs, not in or by itself, but as a conscious decision embraced by some decision makers and rejected by others. The fact that growth can be managed and that persons that initiate ventures do so with many different orientations (Bird, 1988) or propensities towards growth (Sexton & Bowman-Upton, 1986) indicates the need for a significantly different growth model.

OTHER GROWTH MODELS

As a departure from the metamorphic stage growth model, Grenier (1972) proposed a growth model based upon stages of growth in the internal management processes that result from reacting to internal management crises. This model tends to depict the entrepreneur as being forced into stages of growth as the result of organizational management crises beyond the control of the entrepreneur (see Figure 2). In a similar departure, Scott and Richard (1987) combined the crises/change model of Grenier with the metamorphic stage growth model of Churchill and Lewis to develop a

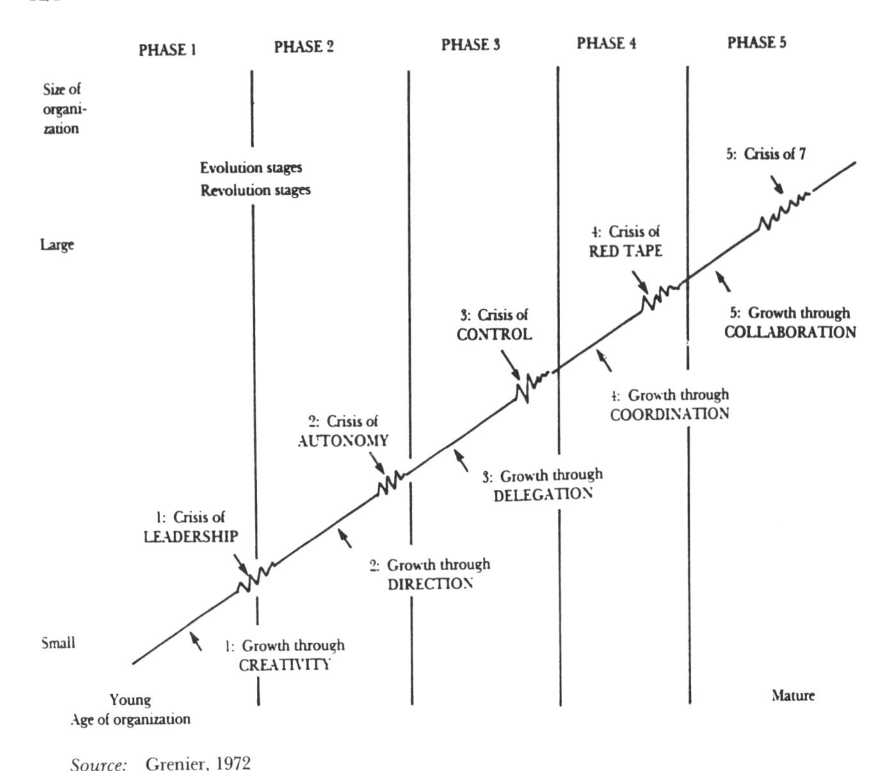

Source: Grenier, 1972

Figure 2. Phases of Growth in a Model or Organizational Development

crisis/metamorphic stage model that incorporates the negative aspects of metamorphic growth stage models with those of a crisis/growth model.

Shanklin and Ryans (1984) have developed an approach to high technology growth that, in some ways, parallels that of Maidique and Hayes (1984). They suggest that growth occurs as the result of a high tech firm making the transition from an innovation driven firm with a supply side marketing approach to a market driven firm with a demand side marketing approach. They further point out that success in one phase does not necessarily mean success in the second phase and that those firms that do make a successful transition do so because they have developed the "right culture."

GROWTH PATTERNS IN EMERGING FIRMS AND A NEW APPROACH TO GROWTH

Reynolds and Freeman (1987), in their studies of new firm start-up, survival and growth, found four patterns of growth that can be described by a combination of initial or first year sales and sales growth during succeeding years. These relationships are shown in Figure 3.

Sexton and Bowman-Upton (1987) developed a 2 x 2 matrix of new venture development based on the CEO's propensity for growth and his/her ability to manage growth. When they combined their studies of the propensity for growth with the empirical studies of Reynolds and Freeman (1987), a similar model emerged in which initial sales were considered as an empirical measure of the decision maker's propensity for growth and subsequent sales were considered as depicting his/her ability to manage growth. The resulting models are shown in Figures 4 and 5.

If growth is a social rather than a natural phenomenon, then growth is the result of a conscious decision on the part of the owner/founder/CEO, and a decision made in conjunction with his/her assessment of his/her ability to manage growth. When the premise is accepted, nongrowth or slow growth is no longer viewed as a failure to grow nor as the inability of the person to manage growth. Instead it is viewed as a combination of the CEO's propensity for and his/her ability to manage growth. Hence the lack of growth is not considered a negative factor. In reality, the growth or nongrowth pattern of the firm is the result of a course of action selected by the CEO.

PROPENSITY FOR GROWTH

What constitutes a propensity for growth? Is it an inborn or developed trait? Are growth-oriented entrepreneurs predestined at birth? The answers to these questions require more than simple yes/no answers. The answers are intertwined in definitional problems and comparative analyses of psychological traits, types or preferences.

To many people, growth is synonymous with entrepreneurship and it is the characteristic that differentiates between a small business and an entrepreneurial firm (Carland, Hoy, Boulton, & Carland, 1984). Others see small business and entrepreneurial firms as being in the same group but with different orientations, which predispose some to make the transition from small to large firms while precluding others from growing (Smith, 1967; Smith & Miner, 1983; Stevenson & Sahlman, 1986). In essence, these studies tend to describe a type of person (opportunistic vs. craftsman, strategic planner vs. non-strategic orientation, promoter vs. trustee) rather than the factors, characteristics or traits that predispose the person to have a particular orientation towards growth.

In general, market factors limit the overall growth that a firm can achieve. Yet some firms achieve higher growth than others under similar market conditions. The differences in organizational growth of new or emerging

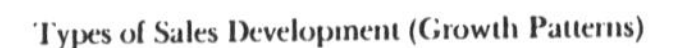

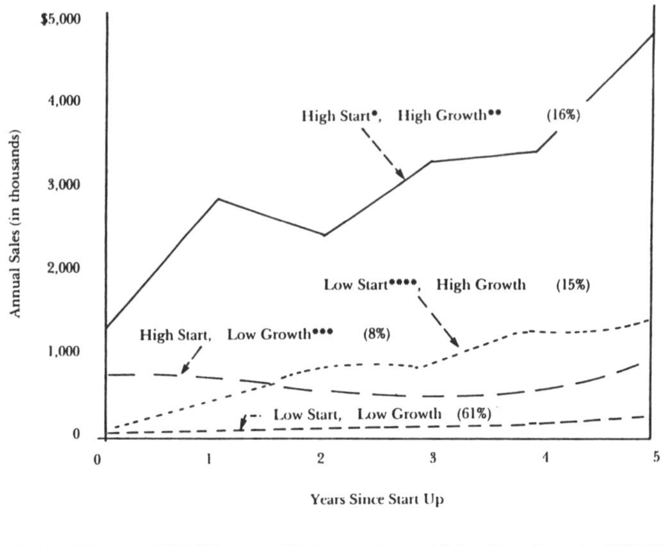

Notes: * Initial Sales over $250,000

**Average Annual Sales Growth over $100,000

***Average Annual Sales Growth under $100,000

****Initial Sales under $250,000

Source: Reynolds and Freeman, 1987

Figure 3. Growth Patterns of New Firms

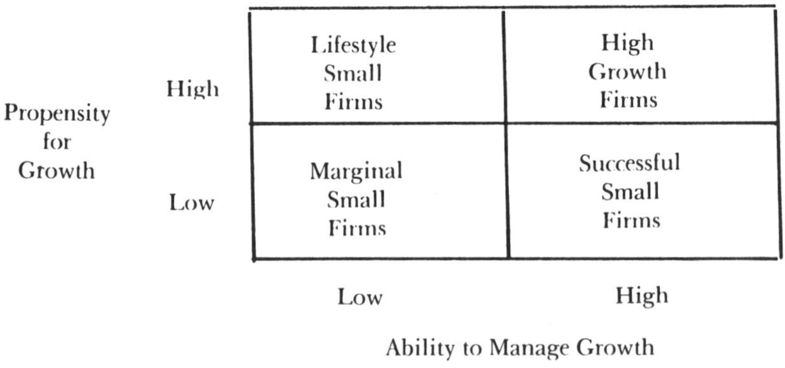

Propensity for Growth		Ability to Manage Growth: Low	Ability to Manage Growth: High
	High	Lifestyle Small Firms	High Growth Firms
	Low	Marginal Small Firms	Successful Small Firms

Source: Sexton and Bowman-Upton (1987)

Figure 4. Relationship between Propensity for Growth and Ability to Manage Growth

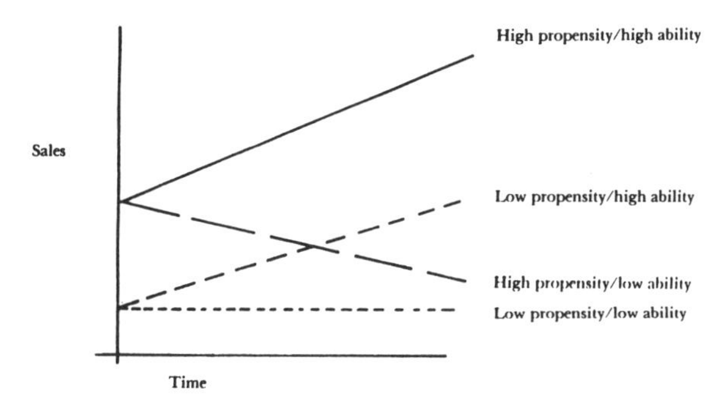

Figure 5. Growth Models of New Ventures As a Function of the CEO's Propensity for Growth and Ability to Manage Growth

firms is related to the entrepreneur's propensity for growth and his/her ability to manage growth. McGuire (1963) expressed the opinion that growth was not spontaneous. He felt that growth was the consequence of decisions related to stimulating demand for a product and than increasing output in response to increased demand. Child (1972) argued that the role of decision makers in the change and performance of the firm had not received the necessary attention in the literature. He described the critical link of the decision makers between the organization and its contextual factors and urged a shift in focus of inquiry from the preoccupation with external constraints and contingencies to the managerial processes that are the basis of the firm's behavior.

Sexton and Bowman-Upton (1985, 1986), in an attempt to find psychological traits unique to entrepreneurs, utilized successful growth oriented entrepreneurs as one component of a sample in their comparative studies of entrepreneurs, managers, small business owners and others. They found that growth oriented entrepreneurs had nine psychological traits with levels of intensity that were significantly higher or lower than the intensities that occurred in others in the comparative studies. According to their studies, growth oriented entrepreneurs scored significantly higher than nongrowth oriented persons ($p < .05$) on scales related to energy level, risk-taking, social adroitness, autonomy, and change. Growth-oriented people also scored significantly lower ($p < .05$) on traits related to conformity, interpersonal affect, harm avoidance and succorance. Additional descriptions of the traits are shown in Table 1. They further concluded that these nine psychological traits constitute a person's propensity for growth or change; that growth is a controllable factor; and that growth is a function of the person's propensity for and ability to manage growth. They argued that having the propensity for growth is, in itself, not sufficient to reach the growth objective of building the organization from its current position to one that is

Table 1. Trait Description of the Personality Scales

Trait	*Description*
Conformity	A low scorer normally refuses to go along with the crowd, is unaffected and unswayed by others' opinion, and is independent in thought and action.
Energy Level	A high scorer is active and spirited, possesses reserves of strength, does not tire easily, and is capable of intense work or recreational activity for long periods of time.
Interpersonal Affect	A lower scorer is emotionally aloof, prefers impersonal to personal relationship, displays little compassion for other peoples' problems, has trouble relating to people, and is emotionally unresponsive to those around him/her.
Risk-taking	A high scorer enjoys gambling and taking a chance, willingly exposes self to situations with uncertain outcomes, enjoys adventure having an element of peril, and is unconcerned with danger.
Social Adroitness	A high scorer is skillful at persuading others to achieve a particular goal, is diplomatic but occasionally may be seen as manipulative of others, and is socially intelligent.
Autonomy	A high scorer tries to break away and may be rebellious when faced with restraints, confinement or restrictions, enjoys being unatttached, free, and not tied to people, places, or obligations.
Change	A high scorer likes new and different experiences, dislikes and avoids routine, may readily change opinions or values in different circumstances, and adapts readily to changes in environment.
Harm Avoidance	A low scorer enjoys exciting activities especially when danger is involved, risks bodily harm, and is not concerned with personal safety.
Succorance	A low scorer does not need the support nor frequently seeks the sympathy, protection, love, advice or reassurance of other people and has difficulty confiding with others.

substantially larger. In addition, the decision maker must have the ability to acquire and manage the necessary resources (regarding capital, raw material, labor, and so on) to meet the growth objectives of the firm. The ability to manage growth functions as a moderating variable to the propensity for growth. The ability to manage growth cannot raise the limits set by the propensity for growth. At a maximum, it can allow the objectives to be achieved, or it can result in some level of accomplishment below the desired growth.

SUMMARY

The contribution of technology based firms to the economic development on a regional and national basis has been a major impetus to growth in the United States. The words "high technology" have become synonymous with growth and change in economic development and many research studies have examined approaches to managing, marketing and financing growth in high technology firms. In this paper the argument is made that growth is a controllable factor and that the initial impetus to growth lies first in the entrepreneur's psychological propensity for growth and second in his/her ability to manage growth. Without a propensity for growth, growth does not occur. Hence, propensity for change or growth becomes a prerequisite for growth in any firm and is even more important in high technology firms where growth opportunities are more prevalent.

REFERENCES

Adizes, I. (1979). Organizational passages: Diagnosing and treating life cycle problems in organizations. *Organizational Dynamics, 8*(1), 3-21.

Arthur Young High Technology Group. (1987). *Biotech 88: Into the marketplace.* San Francisco, CA: The Author.

Birch, D.L. (1987). The booming hidden market. *Inc., 9,* 15.

Bird, B. (1988). Implementing entrepreneurial ideas: The case for intention. *Academy of Management Review, 13*(3), 442-453.

Carland, J.W., Hoy, F., Boulton, W., & Carland, J.W. (1984). Differentiating entrepreneurs from small business owners: A conceptualization. *Academy of Management Review, 9*(2), 354-359.

Cavanagh, R.E., & Clifford, D.K. Jr. (1984). The high-growth potential of midsized companies. *Management Review, 73,* 24.

Child, J. (1972). Organization structure and strategies of control: A replication of the Aston study. *Administrative Science Quarterly, 17,* 163-177.

Churchill, N.C., & Lewis, V.L. (1983). The five stages of business growth. *Harvard Business Review, 61*(3), 31.

Cooper, A.C. (1986). Entrepreneurship and high technology. In D. Sexton & R. Smilor (Eds.), *The art and science of entrepreneurship* (pp. 153-168). Cambridge, MA: Ballinger.

Dale, E. (1952). *Planning and developing the company organization structure.* New York: American Management Association.

Davis, R.C. (1951). *The fundamentals of management.* New York: Harper and Brothers.

Downs, A. (1967). The life cycle of bureaus. In A. Downs (Ed.), *Inside bureaucracy* (pp. 296-309). San Francisco CA: Little, Brown and Rand Corporation.

Drucker, P. (1954). *The practice of management.* New York: Harper and Brothers.

Gardner, D.M. (1986a). *The product life cycle: Its role in marketing strategy.* Discussion Paper, American Marketing Association Faculty Workshop on Marketing Strategy.

———(1986b). *What we don't know about the early stages of the product life cycle: Some preliminary thoughts about the role of the product life cycle and the growth of new enterprise.* Symposium on Research Opportunities at the Marketing/ Entrepreneurship Interface, Chicago, IL.

Gray, B., & Ariss, S. (1985). Politics and strategic change across organizational life cycles. *The Academy of Management Review, 10*(4), 707-723.

Grenier, L.E. (1972). Evolution and revolution as organizations grow. *Harvard Business Review, 50*(4), 37-46.

Hofer, C.W., & Schendel, D. (1978). *Strategy formulation: Analytical concepts.* St. Paul, MN: West Publishing Co.

The Inc. 500. (1987). *Inc. 9,* 75-130.

Kimberly, J.R., & Miles, R.H. (Eds.) (1980). *The organizational life cycle.* San Francisco, CA: Jossey-Bass.

Maidique, M.A., & Hayes, R.H. (1984). *The art of high technology management.* Sloan Management Review, *25,* 18-31.

McGuire, J.W. (1963). *Affecting the growth of manufacturing firms.* Bureau of Business Research, University of Wisconsin, Madison.

Miller, R., & Cote, M. (1985). Growing the next Silicon Valley. *Harvard Business Review, 63*(4), 114-123.

Quinn, R.E., & Cameron, K. (1983). Organizational life cycles and shifting criteria of effectiveness: Some preliminary evidence. *Management Science, 29,* 33-51.

Reynolds, P., & Freeman, S. (1987). *1986 Pennsylvania new firm study.* Washington, DC: The Appalachian Regional Commission.

Reynolds, P., & Miller, B. (1988). *1987 Minnesota new firm study: An exploration of new firms and their economic contributions.* Minneapolis, MN: Center for Urban and Regional Affairs.

Scott, M., & Richard, B. (1987). Five stages of growth in small business. *Long Range Planning, 20,* 45-52.

Shanklin, W.L., & Ryans, J.K. Jr. (1984). *Marketing high technology.* Lexington, MA: Lexington Books.

Sexton, D.L., & Bowman, N. (1985). The entrepreneur: A capable executive and more. *Journal of Business Venturing, 1*(1), 129-140.

Sexton, D.L., & Bowman-Upton, N. (1986). Validation of a personality index: Comparative psychological characteristics analysis of female entrepreneurs, managers, entrepreneurship students and business students. *Frontiers of Entrepreneurship Research,* edited by R. Ronstadt et. al., Wellesley, MA: Babson College.

________(1987). *A growth model of new venture development.* Paper presented at the joint ORSA/TIMS meeting, St. Louis, MO.

Smith, N. (1967). *The entrepreneur and his firm: The relationship between type of man and type of company.* East Lansing, MI: Michigan State University, Bureau of Business and Economic Research.

Smith, N., & Miner, J. (1983). Type of entrepreneur, type of firm and managerial motivation: Implications for organizational life cycle theory. *Strategic Management Journal, 4,* 325-340.

Soukup, W.R., & Cornell, D.G. (1987). Organizational design for high technology firms: The case for controlled instability. Paper presented at the Academy of Management Annual Meeting, New Orleans, LA, August.

Stanworth, M.J.K., & Curran, J. (1976). Growth and the small firm: An alternative view. *Journal of Management Studies, 13,* 95-110.

The State of Small Business: A Report of the President. 1987. Washington, DC: *15,* 39.

Steinmetz, L.L. (1969). Critical stages of small business growth: When they occur and how to survive them. *Business Horizons, 12,* 29-36.

Stevenson, H.H., & Sahlman, W. A. (1986). Importance of entrepreneurship in economic development. In R.D. Hisrich (Ed.), *Entrepreneurship, intrapreneurship and venture capital* (pp. 3-26). Lexington, MA: Lexington Books.

PART III

ORGANIZATIONAL INTERVENTIONS IN HIGH TECHNOLOGY FIRMS

HIGH TECHNOLOGY COMPANIES AND THE MULTIPLE MANAGEMENT APPROACH

Stephen J. Carroll

One result of computer information systems is the appearance of the "information guru." Systems such as manufacturing scheduling have become so complex that only those specialists who work with the computer system daily are able to understand the subtleties of its execution and analyze apparent anomalies in its reports. Hence, management comes to rely heavily on the information guru to interpret and explain some of the most fundamental printed reports. The difficulty with this is that the guru may or may not have the insight into the (in this case) manufacturing process to know whether the information he caresses, massages, feeds and burps is real.

This is not a simple matter to correct. Many of the information specialists are not, by either education or temperament, skilled at critical introspective analysis of data quality. This situation places a burden on the company that has a broad range of products and wide variations, in both volume production and specialized, custom products. In such a case, the information system may come to manage the management. All interdepartmental procedures that impact the manufacturing process tend to be established by the information guru and his minions, whether or not they are the best practices from an overall business standpoint. Important internal procedural decisions tend to be made by lower level staff people who lack the overall business judgement to effect an optimal solution.

One possible solution to this problem lies in the formation of a group chartered specifically to deal with internal practices and procedures. The group should be composed of those management people who are familiar with the overall company objectives and who can call on the lower level specialists for assistance. Much of the initial work of this group will consist of the tedious establishment of paperwork trails, but such details are critical to understanding the overall process. It is in this fashion that the bottlenecks, delays and superfluous reports may be identified. This group can also form a cadre of "specialized generalists" who can understand the whole information system, but who can also maintain the business objectivity necessary to judge its suitability. The efforts of such a management group take time and money, but the cost of having information systems evolve by default or by the earnest machinations of lower level personnel is much greater.

A sure way to hamper the efforts of any such group and to fatally flaw its creativity is to insist upon financial accountability. Even when disguised as long range planning or analysis, financial justification is essentially a short-term process. ROI, ROS, equity turnover and such measurements must necessarily incorporate assumptions and allocations that really have only short range applicability. Many of the basic procedures that such a group must devise have a long range impact on company operations, as they will be called upon to function successfully in new environments as the company grows and its markets change.

—High Technology Company Manager (October, 1987).

MULTIPLE MANAGEMENT AND HIGH TECHNOLOGY COMPANIES

The senior high technology manager quoted above calls for the formulation of a group of "specialized generalists" managers who together have some understanding of the company as a whole to work on the analysis of company systems. Such a group would question current company procedures and reports and would be entrusted to make recommendations for change without having to justify them in quantitative terms. As the manager indicates, this management team would essentially focus on changes that contribute to the long run growth of the company. This manager hopes that this management group might overcome some of the problems of his company arising out of the bureaucratization of some key functions in his company with the advent of new computer information systems and because of the higher degrees of specialization present.

In some companies there *is* such a system in which groups of middle managers put all of the company's operations and products under surveillance with the goal of recommending improvements so that they will function more effectively. This system or approach is called Multiple Management; the purpose of this paper is to describe it and relate it to the needs of high technology companies.

THE NEEDS OF HIGH TECHNOLOGY COMPANIES

The firm described by the manager quoted above is a high technology firm of about 250 employees in an eastern industrial city. The firm's managers all have degrees in electrical engineering and several have graduate degrees as well. The firm's most profitable products are large complex electronic systems designed specifically for each customer's needs and are sold in response to request for bids from these customers. The firm also makes some standardized electronic products in order to provide the firm with a steady source of funds for operating capital. The manufacturing end of the business tends to be more structured and bureaucratic than in the past and similar to manufacturing units in larger companies in other industries. Other units in the firm are more organic (nonbureaucratic) in structure. As the comments of the manager indicate, lately the company has been faced with increasing difficulties in interfunctional cooperation. Arguments and strong disagreements have become more common. Some of the middle managers outside manufacturing believe that this intensified after the company adopted some modern information systems such as MRP, which bureaucratized many of the company's operations. In addition, the inclusion of several different product lines, some with both high and low degrees of task uncertainty, also contributed to some of this firm's present difficulties in meeting customer needs in general and in meeting delivery schedules and customer equipment performance criteria specifically. Increased numbers of management "irritants" have also arisen in recent years. These include system inefficiencies and delays, red tape and interpersonal conflict. Managers in the company are quick to point out, however, that such conflict is not due to internal competitiveness among managers or because of power plays by units as in some of the larger companies in which they previously worked; they are due to narrow specialist perspectives and misunderstandings and the failure of technically trained managers to understand other aspects of the business itself in spite of the company's rather small size.

This company is not unique in its present concerns with interfunctional cooperation system inefficiencies and red tape, and in lack of an overall management perspective and in management skills. Filley (1963) studied firms who grew not by merger or acquisition but by internal expansion. Filley found that such firms moved through several growth stages. The first stage, the entrepreneurial stage, is characterized by personal control, charisma, the exploitation of an innovation and other characteristics typical of this type of firm. Stage two is the dynamic growth stage in which the firm is growing at least 5% a year and the company is devoting almost all of its efforts to exploiting outside opportunities. There is little concern for internal efficiency and effectiveness, and the inputs employed, both human and material, tend to lag substantially behind the resources needed. During this period, many internal management problems arise which are avoided rather than dealt with adequately. These problems often grow in scope and intensity and become more serious.

The increasing revenues and high profits earned during this second period provide some cushion for such inefficiencies but, as Filley (1963) indicates, new competitors are attracted to the industry, putting pressure on the firm to become more efficient by rationalizing operations. Also, new problems are created because the organization's staff is rapidly expanding, with the newer employees exhibiting greater specialization or differentiation along with a lowered emotional commitment to the organization and to the entrepreneur-founder. The new specialists tend not to have the broad perspective of the original employees and also are more likely to be cautious, inclined to stay within their assigned areas of responsibility, and be less favorably oriented toward change (Tosi & Carroll, 1976). Finally, as Filley (1963) reports, the firm arrives at stage three, the rational administration stage, where increased competition causes the firm to become bureaucratic, more cost conscious and less innovative in general.

Some recent research studies on high technology companies, have produced results that conform to the Filley (1963) model. Smith and his colleagues, for example, have studied high technology electronics firms in the State of Washington and in the State of Maryland (Smith, 1986; Smith & Gannon, 1987; Smith, Gannon, Grimm, & Mitchell, 1988; Smith, Mitchell, & Summer, 1985). In the Washington State sample of 27 firms, it was found that such organizations faced increased problems of obtaining political cooperation with increased growth (especially when reaching 500 employees). These political problems of the company replaced the earlier concerns of inefficiencies at moderate levels of size. In a Maryland sample of 28 electronics companies it was found that the better performing companies were more systematically managed than the poorer performing companies (Smith et al., 1988). The better performing companies emphasized, to a greater extent than their poorer performing counterparts, planning and systematic procedures.

This indicates that all firms in a growth situation typically have system inefficiencies, management skill deficiencies and integration difficulties that may seriously undermine their efforts to become established and successful. Betz (1987) has pointed out that a large number of high technology firms fail to survive after a period of initial success and high growth. He attributes this to the firms' failure to correct basic inefficiencies, such as lack of coordination, resulting in an inability to meet customer delivery schedules, and a lack of improvement of first generation products, damaging the firms' credibility and reputation and driving customers to competitors. Rogers and Larsen (1984) described high technology companies in the Silicon Valley area that did not make it and ended up in bankruptcy courts. The primary cause of failure, as they pointed out, was poor management, which was more often the case than other alleged causes such as lack of capital, products with technical deficiencies or poor quality employees. The failure of top management to delegate important responsibilities to lower level personnel also appeared to be a characteristic of failed companies in this area (Rogers & Larsen, 1984). In their descriptions of successful high technology companies such as Intel, Rogers and Larsen (1984) highlighted the importance of constant surveillance of all products and systems and the importance of top management delegation to lower ranks of management.

This suggests that it is not only large established firms that require high management skills, top management delegation, continuous management problem solving and constant improvements in existing systems and products. In fact, high technology firms in all three development stages (Filley, 1963) probably need these process effectiveness characteristics even more than other, larger firms in the absence of "image" credits in the markets they serve. Many of the basic textbooks in management, including several of my own, may have been overly optimistic about the benefits of lack of system and structure in smaller high technology firms. In reality there are very few truly "organic" firms in existence in which there is almost no structuring, there is completely open communications among organizational members, and everyone is in a continuous problem solving mode. In one study of 103 management units in 11 companies (Gillen & Carroll, 1985), it was found that many of the operations in the high technology companies represented in the sample were in fact quite structured, just as in the case cited at the beginning of this paper. The coordination difficulties in such companies with "mixed" structures have been previously described by Tosi and Carroll (1976).

THE MULTIPLE MANAGEMENT APPROACH

Multiple Management is a system of management used by several U.S. and foreign firms, including several high technology firms. This would appear to be a useful approach for helping high technology companies cope with some of their most pressing needs for management development, system improvement and interspecialist cooperation. Although Multiple Management (MM) is not new and has worked very well in a number of firms for many years, the system has received very little publicity over the past twenty years. Few managers and even fewer academics are familiar with it, due to the almost complete absence of research and writing on this approach. One purpose of this paper is to acquaint the reader with this system and to relate its characteristics and effectiveness to the needs of high technology firms. In doing this, some recent research conducted by the author on this system will be described.

The Multiple Management approach was initiated in a dramatic manner in 1932 by the newly appointed president of the McCormick company, Charles P. McCormick (McCormick, 1937, 1949). The company had been founded by Willoughby McCormick in 1889, who developed it into the largest spice company in the world. Willoughby used a rather autocratic leadership style, running the company single-handedly as was the custom of the times. In November 1932, Willoughby unexpectedly died on a trip to New York City, and his nephew Charlie, only 36-years-old at the time, was elevated to the presidency. C.P. McCormick was very different from his uncle in outlook and values. This was clear in his first dramatic actions. He raised the wages of the workforce by 10% even though the company (like most others in 1932) was not in good financial condition. He also had the idea of creating a small group of middle managers close to his own age to act as a *junior board of directors* (a term he coined). This group was to be an elite problem-solving body chosen for their competence and motivation to help improve the company's effectiveness. C.P. McCormick felt that the senior Board of Directors, more resistant to change for a variety of reasons, could not carry out such activities.

In addition to using this group as a source for change and improvement in the company, Charles McCormick felt right at the outset that this type of experience would be excellent training for top management. He believed that experience on the junior board would broaden the individual's overall perspective and would also help develop the analytical, problem solving and interpersonal skills necessary to be an effective manager. He decided to use a rating system for the board members in which they would periodically evaluate each other on their board behavior and performance. These peer ratings were then used to remove the lowest performers from the Board. Participation on the board was a part-time job assignment, and C.P. McCormick provided extra compensation and vacation time for board service. C.P. McCormick felt these incentives would motivate the participants to do as good a job as possible in this secondary work assignment.

C.P. McCormick had an underlying management philosophy on which this system was based (McCormick, 1937, 1949). He was the son of religious missionaries and, for the most part, actually lived abroad in largely undeveloped countries until he was twelve years old. In many ways he was a very humanistic leader, although he was very performance-oriented as well. In his writings and speeches he talked about the necessity of industrial organizations to make the fullest possible use of every organizational member's talents. He was a strong believer in whole workforce participation in organizational functioning. He thought this increased organizational commitment. He also saw employee participation as a means of staving off the collectivist economic philosophies that were quite popular in the 1930s when he started this system. In other words, he saw multiple management as a means of defending the free enterprise system, by providing lower level personnel with the means of participating in the organization's decision-making processes.

We might look at how the system works at McCormick. The MM approach in companies other than McCormick works in much the same way, with some variations. The consistency appears to be due to the fact that the McCormick system was the first; others simply copied it exactly or slightly changed its form to fit their circumstances. The McCormick Company presently has 18 boards, many of which are in acquired companies that had cultures and operating modes that were initially different from McCormick's. All the boards appear to be successful, indicating that top management support and sponsorship can make the system work in a variety of circumstances.

The boards are made up of young managers with at least five to seven years experience in the company. The philosophy is that these managers should learn their own job and their own department's operations first before being elected to an MM board. Board members are typically middle managers, although foremen and even workers sometimes participate. Boards typically meet biweekly, with the members assigned to various projects and teams chosen by the group as a whole at the beginning of term. They work on their assignments when their regular jobs permit. New board members are usually selected by present board members as openings occur. Boards typically drop the lowest ranked members, using various criteria to evaluate its participants. Although those selected are not required to serve, most accept the assignment. Service on the board is limited in length.

Boards range in size from 5 to 20 members. Larger boards can undertake more projects. At McCormick, there are typically two stages of apprenticeship before participants become full board members. Study projects are typically chosen from a list of suggestions presented to the group at the beginning of each term by the members themselves or by other company managers or employees. The meetings of previous Boards also serve as a source of project ideas. Boards usually are free to study whatever they wish and have access to any information they need. The rest of the organization is expected to cooperate, and usually does. When projects are completed, project team recommendations are made in higher management meetings for acceptance and implementation. More than three quarters of the recommendations are accepted but sometimes implementation is delayed for a variety of reasons.

The Multiple Management system at McCormick has evolved over the years as a result of considerable study. Because C.P. McCormick and the Multiple Management system believe that nothing is sacred and everything should be under constant surveillance for purposes of improvement, the MM system itself has been periodically studied by both outside investigators (such as myself) and by internal MM boards. As a result of these investigations, the operation of the system at McCormick is now different from in the past. For example, at one time the MM Junior Boards ran the employee suggestion systems of the company, in addition to developing their own suggestions. Now only some boards carry out this activity. In the past, the Boards also managed a formalized mentoring system in addition to their other responsibilities. The number of boards has changed as well as the procedures they follow, in addition to changes in the scope or domain of the boards' activities. There are now 18 boards at McCormick. Some have fairly general activities while others confine themselves mostly to sales issues or production problems.

As previously indicated, some MM Boards at McCormick and in other companies supervise the employee suggestion system and some do not. While one might think this would be a worthwhile activity, most of the McCormick managers interviewed whose boards did evaluate employee suggestions indicated that results were rather meager. Since another procedure is used for obtaining employee suggestions on benefits and working conditions (surveys and sounding boards), few performance-enhancing suggestions are received. This is quite typical of most employee suggestion systems. Of course it is possible that the present relationship between the employee suggestion system and the board system has not been designed and administered in an optimal manner.

One McCormick manager said in an interview that one of the reasons the board system made a real contribution to the Company was that these boards could work on improvements that the regular operating units did not have time to investigate. Well-known management scholars have often remarked that it is very difficult for an existing operating organization to be innovative. According to a number of writers on this topic, an organization for innovation, separate from the organization responsible for operating results, must be created in order for innovation to occur. For example, Drucker (1986) describes when he was a consultant at General Electric in the 1950s and how he was wrong in assuming that general managers could be responsible for both current and future operations. He goes on to say that this system did not work at all because the general managers responsible for innovation were actually being rewarded for the present performance; any long term innovation expenditures would not promise a return for some time in the future. Similarly, Galbraith (1982) and Kanter (1985) also point out the importance of separating an innovative organization from an operating organization.

With the board system, McCormick and other companies, in effect, do have two such organizations simultaneously, and the MM board members work in both organizations at different times. The MM board system is the innovative organization, separate from the operating organization, and is not concerned with current deadlines or operating issues. It can work on the process of change and revitalization, which many argue is very much needed in U.S. companies today. Galbraith (1982) pointed out that one problem with having two such organizations is the implementation of the ideas in the operating organization. Some way must be found to bridge the gap so that NIH (not investigated here) factors are not operating. The fact that the board members have high credibility and good standing in the operating organization certainly facilitates this implementation process. The fact that the recommendations must go the highest level in the organization for authorization and endorsement helps as well. Finally, these transitions are facilitated by the typical habit of the boards in co-opting individuals in the affected segments of the unit by getting them personally involved in the project.

Since its beginnings at McCormick, the Multiple Management system has spread to other companies. Many other companies adopted this system after it was publicized by C.P. McCormick (1937, 1949). In addition, Craf (1958) presents a summary of an evaluation of Multiple Management in twenty-one companies. The exact number of companies using this system is unknown: Craf (1958) estimated the number of companies using the system at about 100 in the 1950s; C.P. McCormick (1949) claimed many more. This investigator followed up some of Craf's companies and found the system was still in use in such companies as Motorola and Lincoln Electric, but that it had disappeared in other companies for a variety of reasons. It seems about forty organizations use the system at present.

THE EFFECTIVENESS OF THE MULTIPLE MANAGEMENT APPROACH

The effectiveness of any system of management such as Multiple Management, or a particular inventory control, purchasing, or human resource management system must be evaluated on the basis of attaining its objectives. The objectives or purposes of the MM system were stated by C.P. McCormick (1937, p. 4) when he created the first Junior Board of Directors in 1932:

> I called together seventeen of our younger men who appeared to be promising. They were assistant department managers and others who had been taking a special interest in their work. At a preliminary meeting I carefully explained that our purpose was not to supersede the judgement of the men who had made the business a success but to supplement that judgement with the energizing power of new ideas. I also explained that the greatest value anticipated was the training of executives.

Later statements by company officials also support these early goals of McCormick and sometimes add a third goal—that of improving cross functional relationships.

Of course, any system of management must ultimately contribute to bottom line results for the organization as a whole. But it is almost impossible to determine the financial impact of one system on a company's organization, as indicated by the high technology manager quoted at the beginning of this paper. One investigator of Multiple Management at McCormick did a careful statistical analysis of McCormick's success relative to its competitors in economic terms for the years 1968 to 1978 (Stiehl, 1980). He found that McCormick had the best record for controlled and sustained growth compared to its seven competitors. However, McCormick managers themselves are fairly consistent in believing that the MM system is only one of a whole host of factors responsible for the economic performance of the company over the years.

After some thought, it seemed that the best way to start to evaluate the effectiveness of the MM system was through examining the innovations themselves to see if they were of value to the company and by interviewing the managerial participants on the board to determine if they had improved their management skills as a result of their board participation. This investigator was fortunate to find several other surveys, in addition to his own, to assist with this task, making it unnecessary to study a large group of company managers. So far this investigator's interviews have been conducted with about 30 present and past board members, including most of the present top managers at McCormick. These interview results were compared to a survey by MM boards themselves in 1977 and 1979, along with an interview survey of the earliest board in 1936. Finally this investigator was given some data from a survey conducted with more than 600 completed questionnaires from McCormick managers, that were still being tabulated. The findings from the interview studies conform to the results of this questionnaire study.

There is certainly evidence of the effectiveness of the MM board system in the comments of the participants themselves. In all of the interview studies there was virtually universal agreement that the system does produce a wide variety of useful management development experiences and useful innovations or changes in organizational systems and products. There is very high consistency in the reports of the participants themselves in the types of management development benefits received. They fall into the three categories identified in the 1979 self-study of McCormick MM participants: (1) refinement of management skills (time management, problem solving, communication, etc.); (2) exposure to personalities and operations in other parts of the company (e.g., organization knowledge); and (3) cultivation of cooperative relationships that facilitates daily job performance. This last benefit leads to another important present goal of the MM system — that of improving interfunctional cooperation. It results from the fact that the typical MM boards are made up of middle level managers from different functional areas who learn a great deal about one other and one other's operations through close association over a considerable period of time. In more than 95% of the interviews examined, at least one of these factors was mentioned. In more than 70 percent of the interviews, all three of these factors were mentioned. Some examples of comments in interviews include:

> Being on the board gave me an opportunity to learn management. Now I can talk to operating personnel around the company and understand what they're saying.
>
> Some of the main gains to me from being on the board was learning to organize your work because you have another regular job. Board membership is terribly time consuming. Also you learn presentation skills which are important. You also learn the essentials of organizational politics. There are politics everywhere. You learn to juggle many balls at once which you must know how to do in management.
>
> You learn by making mistakes. On the boards your mistakes are not costly to the company or to you as on your regular job later.
>
> As far as the benefits of being on the board I can see it was good for getting known to top management. What I didn't see at first was the broadening. Of seeing how other people and other parts of the company and other disciplines operate. Also it was real management training. You developed skills in diagnosis, in decision making and in making presentations. Also you learn how to counsel younger people when you are more senior on the board. The board helps you actually get a lot of authority. Also you learn time management. How to prioritize your assignments and activities.
>
> When I first came here I knew nothing about how organizations really operate since I was trained in a scientific specialty. I found out on the board about organization politics. That life in an organization is a type of game you have to learn how to play. It isn't just telling others the truth that you discover through some rational means as they tell you in science classes. Success in an organization even for a scientist involves preselling others first. It involves learning how to sell yourself and your projects. It means learning presentation and other communication skills. It also means learning about your own strengths and weaknesses. These are the things that I learned through my board experiences. This type of management training is especially needed by technical personnel like myself.

Comments from managers in other organizations, using the system authored by this investigation, were quite similar. For example, some comments from managers in a technological organization employing many scientists and engineers:

> Being on the council means management training without the risks on a regular line job. This is because you can fail and it doesn't have any bad consequences for others.
>
> The training on the Board was especially useful to me with my training as an engineer. Engineers in college receive no training at all in being a member of a group and learning how to be a good team player and how to motivate others.
>
> In an organization filled with technical specialists, it is very useful to have an opportunity to learn about the organization as a whole especially the support departments such as purchasing and personnel.

The interviews with these other organizations using the system in 1988 showed great consistency with the comments made by McCormick managers. This is similar to the findings of Craf (1957), who compared twenty-one companies using the system, and found great similarity in characteristics of the system employed and also in managerial reactions to the system. The interviews, as well as the large questionnaire study of the system carried out by an MM project team, revealed that ratings of the effectiveness of the system for management development were higher and less variable than ratings of the system's effectiveness for producing innovation.

A STUDY OF THE VALUE OF MM BOARD PROJECTS

Since the interviews with both McCormick and other organization MM Board members had revealed high variability in the quality of MM board suggestions, this investigator felt it would be worthwhile to explore some of the reasons for this variability. Such information might then be used to reduce the variability in the quality of projects, and might make the MM boards as effective for producing quality projects as they were for management development and reduction of the "tunnel-vision" problem among specialists previously described by Kanter (1984).

A questionnaire designed to explore this issue was sent to 137 past and present MM board members in McCormick and in several of its subsidiary companies. Seventy-eight questionnaires were returned. In each questionnaire, the manager described one of the most successful projects he or she had witnessed and the name of one of the least successful projects observed while an MM board member. Each manager surveyed was then asked to distribute 100 points among seventeen items listed on the questionnaire according to their perceived contribution to this project's success or lack of success. The factors listed were previously obtained in the interview study with McCormick managers. Respondents were also allowed to write in any item not mentioned and weight it accordingly as well. However, there were very few write-in responses.

First, an analysis was made of project focus and its relationship to project success. This was done by reading the descriptions of the most successful and least successful projects to determine if the subject or focus of the project, such as product improvement, technological improvement or administrative procedure improvement, was related to a project being classified among the best or worst projects ever done. A preliminary analysis indicated no consistent tendency for certain types of projects to be classified as more successful as opposed to less successful projects. For example, several projects having to do with product analyses were listed among the best projects ever witnessed by the respondents, but some product analyses were among the worst. The same was true of studies of company technological processes and administrative procedures. Even projects focusing on the company's competitive strategies for different lines of business were listed among both the best and worst projects. This shifted attention to the factors related to a project's success regardless of subject or focus of the project.

The interviews with the McCormick managers indicated that there were several factors that would probably be related to a project's success. One factor was top management's interest in a project and thus its motivation to accept and implement a project's recommendations. Another was the interest of the project team and the overall MM board itself in the subject matter or issue involved with a particular project. Another factor would be simply the time and energy invested into choosing and working on the project. Less time and energy applied to a project would result in a lower quality project and thus one that would be less effective and successful. Still another factor would be the skills of the project team. These might include technical skills, problem solving skills, persuasive skills (for selling the project to top management) and the appropriate combination of skills on a particular project team. The experience of the project team members in carrying out similar projects might also be associated to how effectively this particular behavioral technology is used. Another factor thought to be of possible influence on the project was degree of cooperation. This could include cooperation among the project team members (internal conflict might waste energy), and cooperation with the rest of the organization in providing the project team with data or other needed assistance. These factors are shown in Figure 1. The five overall factors were then converted into seventeen specific items, which were then randomly mixed in order of presentation to the managers in the questionnaire.

Table 1 presents the weights given to all seventeen items or factor elements in terms of their perceived contribution to the success or lack of success of

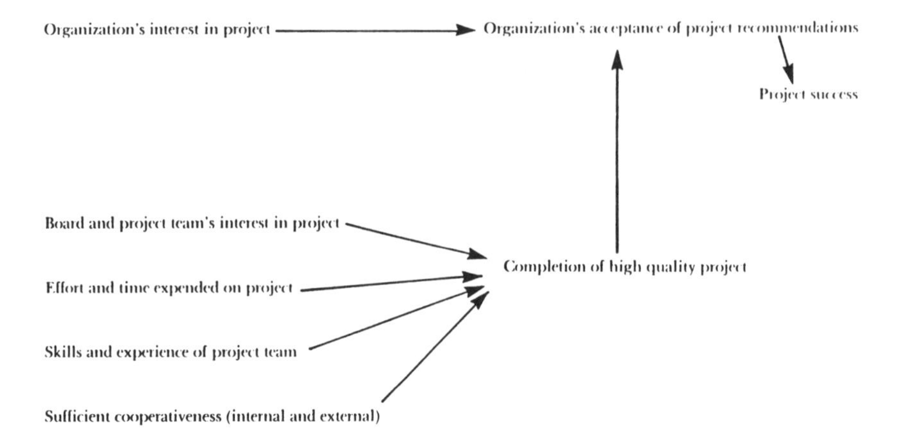

Figure 1. Factors Related to Project Success

particular MM board projects. Table 1 shows that the most important factor in perceived project success was the relationship of the project to the company's present interests and issues. If a project is to be a success the MM board must only choose projects that will be of high interest and relevance to top management in the company. Obviously preproject selection contacts with top management are needed to accomplish this.The procedures for doing this varies from one MM board to another at McCormick. Table 1 indicates that there are several other factors that appear to be strongly associated with project success. The project team's degree of interest in the issue being studied is very important, having a significant impact on the project team's motivation levels.

The motivation of the project team is especially important under Multiple Management because the boards are made up of volunteers who carry these activities in addition to their regular job duties—often in the evenings and on weekends. It would appear from this analysis that the system for selecting projects should include making a list of projects of interest to the members which is then matched against a list of problem areas and issues from top management. There are a number of specific ways to accomplish this. The next most critical factor in the success of a project is the managerial ability of the project supervisor. He or she must insure that deadlines for various parts of the project plan are accomplished on time and that important steps are not overlooked, in addition to coordinating the activities of various members of the project team. A careful selection of the project team chair is important. In terms of project success, it would appear that obtainingcooperation from the rest of the organization and using goals and deadlines are also somewhat important to project success. The rest of the

Table 1. Average Weight Given to Various Elements as Determinants of the Degree to Which Multiple Management Board Projects Were Successful or not Successful in Relative Terms

	Most Successful Projects		*Least Successful Projects*	
Potential Success Elements in Project Success	*Average Weight*[a]	*Rank*[b]	*Average Weight*[a]	*Rank*[b]
Degree of company interest in project	19.3	1	16.3	1
Degree of project team interest in issue	12.6	2	6.9	4
Degree of relevance to current organizational issues	12.6	2	6.9	4
Degree to which project checked out first with top management	7.4	4	7.7	3
Managerial ability of project team chairman	6.3	5	6.8	5
Cooperation from rest of organization on project	5.8	6	4.1	10
Amount of care in selecting project	5.7	7	6.6	6
Necessary technical skill on project team	4.2	9.5	4.7	8
Degree of problem solving skill on project team	4.1	11.5	3.4	12.5
Amount of time spent on project	4.1	11.5	3.4	12.5
Amount of MM experience of project team members	3.1	13	3.0	14
Degree of cooperativeness on project team	1.8	14.5	1.3	17
Degree of appropriate mix of talents on project team	1.8	14.5	3.4	12.5
Degree of interest of MM board chairman on project	1.1	16	1.9	15.5
Degree of persuasive skills on project team	.6	17	1.9	15.5

Notes: [a]Weight given on a 1 to 100 point scale on perceived contribution to degree of success.
[b]Rank on the basis of average weight given by respondents on perceived contribution to degree of success.

elements listed are far less significant. It should be noted that these element weights were not obvious in the interviews with the MM board members. The managers interviewed tended to just list elements without specifying degrees of relative importance. Finally, it appears that there are no significant differences in the factors related to project success versus project failure. Initially it was felt that the factors related to high project success may be different from those related to low project success.

MULTIPLE MANAGEMENT IN OTHER COMPANIES

While most of the data in the form of interviews and questionnaire results were collected in the McCormick company and its subsidiaries, this investigator also contacted eight other organizations using this approach. These organizations completed a long questionnaire on their system—its history, unique characteristics and perceived effectiveness. In addition, some telephone and in-person interviews were carried out with managers in some of these other organizations. For example, one organization that was studied in more depth was a government-military organization involved in advanced weapons research that employed many scientists and engineers. In this government high technology organization, a number of interviews were carried out with present and former board members, and the organization's archives were examined.

Another organization with the MM Board system is Cutter Laboratories, a biotechnology company in California. Cutter, which classifies itself as a high technology company and has used this system for more than thirty years, reports great satisfaction with the system and plans to continue its use. At Cutter, the board is viewed as a means of achieving management development outcomes and getting managers just below the middle level to participate in managing the company, as well as producing valuable ideas. Cutter's board operates in the same way as McCormick's system. Every six months some members are rated off and new ones are rated on. Presently, the board consists of nine members. They have complete authority to choose their own projects, investigate them and make recommendations to top management. The board is currently working on several projects designed to improve employee morale and keep employees with the company at a time when good employees are not easy to attract and retain.

In general, this investigation indicated that the MM system in these other organizations produced results similar to those at McCormick. All of these organizations had successfully used this approach to management for a number of years. Some of them differed in the emphasis that they gave to the various outcomes of this system. For example, most gave the highest priority to MM as a means of developing general management skills and broadening the perspectives of specialists. Especially important was broadening the outlook of managers in such technically advanced organizations as the government weapons facility and a biological research company in California. In one company, a southern food products company, the MM board was comprised of equal numbers of workers and managers, since the primary purpose of this board was to break down worker-management barriers and give workers a chance to help manage the company. All of the organizations indicated the primary value of the approach was in management development and improving interfunctional cooperation. They agreed with the McCormick managers surveyed that the value of the ideas from the boards in terms of system and product improvements was quite variable.

MULTIPLE MANAGEMENT AND CURRENT MANAGEMENT PRESCRIPTIONS FOR IMPROVEMENT

A number of recent and popular management books have expressed the idea that companies need to pay more attention to the systematic and incremental improvement of existing systems and products, development of higher basic management skills, and the break down of internal unit barriers to achieve higher operational synergies in order to be more competitive. Peters (1987) talks about the importance of having middle managers play a newer role in organizations—that of barrier bashers, among other things. Kanter (1984) describes how six especially successful companies manage innovation and change. She talks about departments as fortresses and how cross-functional middle management teams are an important ingredient in creating successful change. Drucker (1985) indicates that all companies must have specific organizational systems and procedures for producing the needed innovations.

Perhaps the greatest difficulty faced in high technology companies is the inordinate amount of attention paid to technical factors and considerations relative to managerial concerns and issues. This is partly because the employees and (especially) managers and professionals in such firms are technically oriented by training and by choice; they are most interested in and most sensitive to technical and scientific issues. Management perspectives, however, are more pragmatic and economic. How can we afford something we need and want? How can we get everyone working together in harmony? How can we best utilize our human and material resources so as to best achieve a difference between outputs and inputs? What exactly is the customer's perspective in examining our products and services? Technically trained individuals especially need to become more sensitive to such questions and to become more aware of some alternative ways of approaching and solving such problems. Being forced to confront such problems in an organization of members with quite different backgrounds and modes of thinking is one way of dealing effectively with this issue.

The Multiple Management system helps to accomplish the three key organizational needs, which, it has been argued, are especially important in high technology companies: management development; incremental system and product improvement; and interfunctional cooperation. The MM system has operated successfully in some U.S. companies for many years (55 years in the case of McCormick and 45 years in the case of Lincoln Electric). It has worked well in a variety of industries and in companies with different cultures, histories and experiences. On the other hand, there are certain prerequisites for its success, just as there are with all management systems. Some research has been described that may help to identify such prerequisites. Multiple Management is a system that has, by and large, faded from the collective consciousness of management practitioners and scholars but, as this paper indicates, should be revived. It seems to represent a viable solution to some of the most pressing problems of U.S. companies, especially high technology companies.

REFERENCES

Betz, F. (1987). *Managing technology*. Englewood Cliffs, NJ: Prentice-Hall.

Craf, J.R. (1958). *Junior boards of directors*. New York: Harper & Bros.

Drucker, P.F. (1986). *Innovation and entrepreneurship*. New York: Harper & Row.

Filley, A.C. (1963). *A theory of small business and divisional growth*. Unpublished doctoral dissertation, Ohio State University.

Galbraith, J. (1982). Designing the innovating organization. *Organizational Dynamics, 10*(1), 5-25.

Gillen, D.J., & Carroll, S.J. (1985). Relationship of managerial ability to unit effectiveness in organic versus mechanistic units. *Journal of Management Studies, 22*(6), 668-676.

Kanter, R.M. (1984). *The change masters*. New York: Simon & Schuster.

________(1985). Supporting innovation and venture development in established companies. *Journal of Business Venturing, 1*(1), 47-60.

McCormick, C.P. (1937). *Multiple management*. New York: Harper & Bros.

________(1949). *The power of people*. New York: Harper & Bros.

Peters, T. (1987). *Thriving on chaos*. New York: Alfred Knopf.

Riggs, H.R. (1983). *Managing high technology companies*. Belmont, CA: Lifetime Learning Publications.

Rogers, E.M., & Larsen, J.K. (1984). *Silicon valley: Fever-growth of high technology culture*. New York: Basic Books.

Smith, K.G. (1986). Organizational maturation and professional management. *Futures, 18*(5), 671-680.

Smith, K.G., & Gannon, M.G. (1987). Organizational effectiveness in entrepreneurial and professionally managed firms. *Journal of Small Business Management, 25*(3), 14-21.

Smith, K.G., Mitchell T.R., & Summer, C.E. (1985). Top level management priorities in different stages of the organizational life cycle. *Academy of Management Journal, 28*(4), 779-820.

Smith, K.G., Gannon, M.J., Grimm, C., & Mitchell, T.R. (1988). Decision making behavior in smaller entrepreneurial and larger professionally managed firms. *Journal of Business Venturing, 3*(3), 223-232.

Stiehl, R.T., III. (1980). *Innovations and cooperation through boards for the middle manager*. Unpublished thesis, Department of Anthropology, Graduate Faculty of the University of Virginia.

Tosi, H.L., & Carroll, S.J. (1976). *Management: Contingencies, structure and process*. Chicago: St. Clair Press.

HOW TO DEVELOP A HIGH TECH FIRM INTO A HIGH PERFORMANCE ORGANIZATION

D.D. Warrick

High tech firms are playing an increasingly vital role in our world economy and in contributing to our knowledge about adapting organizations to a rapidly changing environment. Some, such as Hewlett-Packard, IBM, Digital and Apple are emerging as pace setters among the excellent organizations studied, imitated by others and heralded in business books and journals. However, all is not rosy for the high tech industry. It has proven to be a highly competitive, volatile and constantly changing industry that has enjoyed little stability and has witnessed organizations rising and falling at a dizzying rate. These conditions have placed a premium on two key skills: (1) *the ability to successfully manage change*; and (2) *the ability to develop a High Performance Organization.* Lack of attention to these two critical skills represents a sure formula for eventual mediocrity and possible obsolescence for high tech firms.

MANAGING CHANGE IN HIGH TECH ORGANIZATIONS

We now have the technology to successfully manage change and develop high performance organizations. The field of *Organization Development* is devoted to the study of managing organization change and development (for example, see French & Bell, 1984; Burke, 1987; Huse & Cummings, 1985; Warrick, 1984). In addition, many recent books have described various approaches to developing high performance organizations (for example, see Kilmann, 1985; Tichy, 1983; Kanter, 1983; Weisbord, 1987; Peters & Waterman, 1982; Peters & Austin, 1985; Peters, 1987).

How applicable is current change technology to high tech organizations? Certainly, most sound change principles will work with any organization. However, Table 1 shows several unique characteristics of high tech firms that have important implications for designing change strategies.

Dynamic Environment

Perhaps the most prominent unique characteristic of high tech organizations is the dynamic environment in which they operate. The intense competition and struggle for stability accentuate the need for having a change strategy and for using accelerated methods for accomplishing change. New methods need to be explored for accomplishing lasting change as quickly as possible.

The constantly changing environment also creates an interesting paradox. While change is somewhat of a way of life, additional change may be resisted for fear that it will make an already chaotic situation more chaotic. This suggests the need to provide incentives for committing to change and to plan ways to build commitment to the change program. Commitment building activities should be targeted for the top person in the organization, the top management team, and others who could have a significant influence on the outcome of the program.

Technical Emphasis

The strong technical emphasis in high tech organizations and desire by many technical professionals to "be left alone to do their job," often relegates planned change to a low priority or no priority. Likewise, although managing change may significantly influence the success or survival of the organization, widespread interest in change programs may range from minimal to nonexistent. These realities suggest that change programs need to be "championed" by top management so employees at all levels will grasp the importance of the program and make the necessary commitments to make the program work. In addition, improvements made as a result of the program need to be visible and communicated to maintain momentum and interest.

Table 1. Unique Characteristics of High Tech Organizations and Their Implications for Managing Change

High Tech Characteristics	*Implications for Managing Change*
Dynamic Environment (Constantly Changing, Volatile, Unstable)	• Change efforts and methods need to be accelerated. • The constant change may result in resistance to additional change. Therefore, it is important to have a strategy for building commitment to the change program.
Technical Emphasis	• Change efforts must be conceptually clear and supported by top management so there is a buy-in to a nontechnical program. • Results must be visible and communicated to maintain momentum.
Underdeveloped Concept Of Leadership	• Top level managers need to be coached on the importance of leadership and unity at the top and developed into a high performance top management team committed to championing change and shaping organization culture.
Loose Structure (Relatively Flat, Unclear Boundaries)	• Structuring the organization for results, clarity, and flexibility is likely to emerge as a central issue.
Emphasis On Teamwork	• Cooperation and collaboration need to be developed within and between groups and between professionals and technicians.
Project Focus	• Managers and project leaders need to be trained how to quickly develop high performing continuing and temporary teams
Inexperienced Managers	• Management development needs to be emphasized with a focus on self-management and on empowering and enabling employees rather than managing in a traditional sense.
Employee Needs For Ownership And Growth	• Employees at all levels of the organization need to be involved in the change program and activities need to be planned that appeal to the growth needs of employees and that give employees an incentive to stay involved.

Underdeveloped Concept Of Leadership

The "excellence" literature and recent books and articles on leadership (for example, see Bennis & Nanus, 1985; Carlzon, 1987; Kouzes & Posner, 1987; *New Management,* 1988; Tichy & Devanna, 1986) suggest that *leadership* is the most critical variable in developing high performance organizations. Nations and organizations rise and fall based on the quality of their leaders. High tech organizations often have a strong leader at the top who is respected for his or her technical expertise and entrepreneurial zeal. However, such leaders may not have a broad understanding of leadership and the need to develop a top management team committed to leading the organization by providing *vision, direction, and inspiration.* Therefore, developing the top management team into a leadership team will likely be a major focus of a change program in high tech organizations.

Loose Structure

High tech organizations are often loosely structured with fuzzy and multiple reporting relationships and responsibilities, a relatively flat structure, and unclear reward and evaluation systems. The lack of attention to structural issues often results in employees getting mixed signals about what management proclaims and what, in fact, the system rewards. For example, teamwork may be emphasized while all rewards are in fact based on individual effort. The strength of a loose structure is flexibility. The weakness is the risk of lack of direction and confusion about what is expected and valued, who does what, and knowing where you stand.

In designing organizations for results, the goal should be to:

1. align the structure, and particularly the *Performance Review* and *Reward Systems,* to be consistent with the mission, philosophy, and goals of the organization;
2. assure that the right people are in the right place with clear responsibilities;
3. assure that skilled leaders and managers are placed in key positions;
4. develop a lean, flexible, nonbureaucratic, functional and results oriented structure that encourages self-management, high productivity and innovation.

Most high tech organizations will need to be restructured to be truly designed for results. Managers will also need to develop skills in managing the process of restructuring since restructuring will become commonplace in a dynamic environment.

Emphasis On Teamwork And Project Focus

The need for teamwork to accomplish complex tasks and the project focus of high tech organizations both have important implications for managing change. Teamwork needs to be developed both within and between groups. Also, the frequent use of temporary project teams suggests the need to train managers and project leaders how to quickly develop teams into high performing teams. It is a rare organization that has trained its managers and project leaders how to develop teams. And yet, the lack of training in this important skill often results in teams floundering and experiencing low performance and morale and minimal creativity in solving complex problems.

Inexperienced Managers

In high tech organizations, many managers are simply technical professionals with management titles. They are often thrust into management positions with little if any management training. It is curious to consider how unthinkable it would be to place a person in a technical position with no prior training. And yet, it is commonplace to place people into management positions who have had no training in management. If high tech organizations are to succeed in an environment characterized by intense global competition, they cannot escape the need for developing excellent managers. Thus, *management development* should be a major focus of change programs in high tech organizations. The unique characteristics of high tech organizations suggests the development of managers who: (1) can find the balance, depending on their level in the organization, between the use of technical, management and leadership skills; (2) emphasize creating a work environment that encourages self-management; and (3) see their primary role as empowering and enabling employees to perform at their best.

Employee Needs For Ownership And Growth

Change programs often never penetrate the organization beyond management. Such programs are likely to be short lived in high tech organizations where employees have strong needs for ownership and personal growth before committing to change. It is important to involve the appropriate employees at *all levels of the organization* in managing changes to increase ownership in the changes resulting from the program. In addition, when involving employees, it is helpful to include some training so employees will have an immediate personal benefit and will look forward to nontechnical sessions designed to strengthen the organization.

DEVELOPING A HIGH PERFORMANCE ORGANIZATION

The wealth of information now available on the "excellent" organizations has made it possible to study why these organizations are able to perform at such high levels. We read of organizations such as Scandinavian Airlines being on the verge of losing $20 million and one year later earning $54 million. In the automobile industry, we find that Ford Motor Company is able to make almost twice the profits of General Motors on half the sales. In the steel industry, we read of Chaparral Steel taking 1.7 man hours to produce a ton of steel while the industry average is 6 hours. Examples of many other organizations that are producing from two to ten times the industry average or that provide superior service and quality products while maintaining high productivity are readily available.

Why aren't more high tech organizations taking advantage of the technology presently available for managing change? Perhaps the foremost reason is the lack of awareness that the technology exists. However, another dilemma is integrating the available knowledge into a change strategy that also addresses the unique characteristics of high tech organizations.

The model in Figure 1 and the step-by-step strategy for developing a high performance high tech organization shown in Table 2 are attempts to integrate sound change principles drawn from the field of organization development with what has been learned from the excellent organizations and with the unique characteristics of high tech organizations.

The model recognizes three stages of change: (1) Preparation; (2) Implementation; and (3) Transition. Unless all three stages are understood and addressed, even the most well intended and needed changes will enjoy only temporary success or will fail. The heart of a change strategy designed to develop a high performance organization is to continuously strengthen and fine tune five target areas:

1. Leadership;
2. Organization Structure;
3. Teamwork;
4. Management;
5. Personal Utilization, Development, and Involvement.

IMPLEMENTING A CHANGE PROGRAM

Change efforts that have considerable potential often prove disappointing. This may happen because the organization is unrealistically looking for "quick fix" technical or training solutions or is solving the wrong problems. However, the primary reason is a lack of understanding of the change process. The change program described below shows how change can be accomplished using the change model in Figure 1 and the step-by-step strategy presented in Table 2. While the activities described may appear to be sequential, in actual practice, many occur simultaneously and are overlapping.

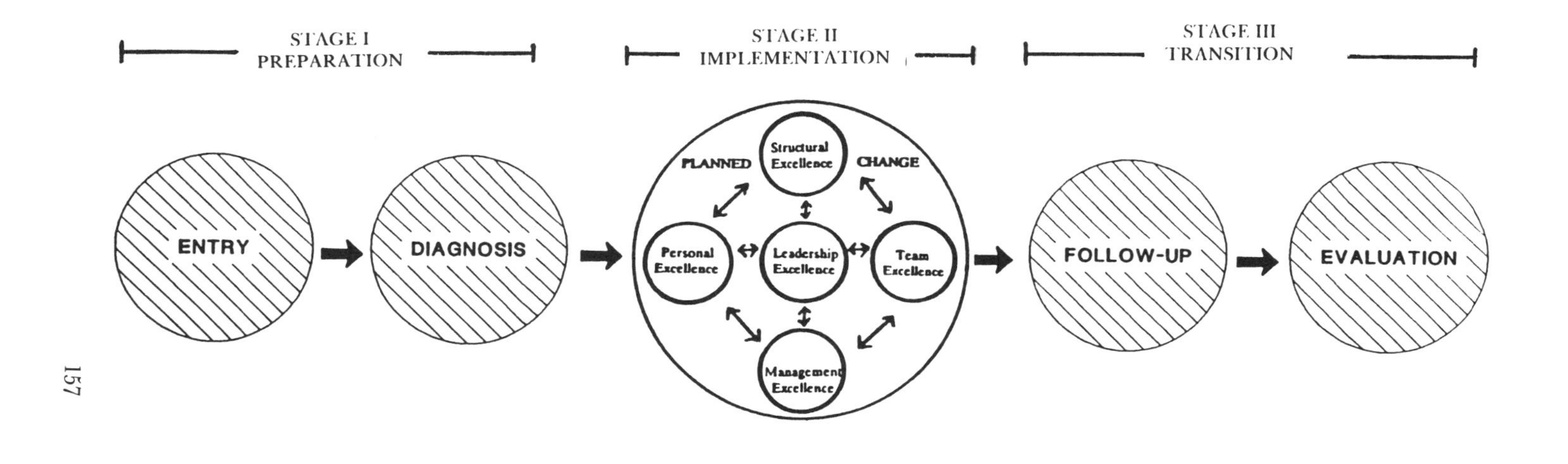

Figure 1. Planned Change Process

Stage I—Preparation

Stage I consists of an *Entry* component and *Diagnosis* component. The purpose of the Entry component is to explore organization needs, consider alternatives for addressing the needs and build commitment to the agreed upon change strategy from those persons who can most influence the outcome of change efforts. Commitment is demonstrated when these key people show support for the program, are willing to be involved and are willing to champion change efforts. When the key stakeholders are committed, change will almost always succeed. Without their commitment, change efforts will achieve moderate success or eventually will fail.

The Diagnosis component usually consists of a survey administered to employees and a cross-section of interviews with a sampling of employees. This phase provides employees throughout the organization with an opportunity to contribute their ideas and prepares them to expect change.

Stage II—Implementation

Actual changes are implemented in Stage II. It includes a focus on *Planned Change* that relies on individual, group and systems interventions directed primarily at improving *Leadership, Organization Structure, Teamwork, Management, and Personal Utilization, Development and Involvement.* Although traditional change models begin interventions with individual training followed by team building and eventually attention to organization wide issues, the planned change approach described in Table 2 works in reverse in recognition of the need to accelerate the change process in high tech organizations. It begins conceptually with organization-wide change followed by team change and individual change.

The Planned Change phase begins with interventions with the Leadership Team (top management team) designed to consider organization wide issues such as clarifying the mission, philosophy (values) and goals of the organization; defining the role of the top management team in providing leadership and championing change efforts; and using the results of the organization assessment to target needed changes. Then, a strategy is developed to involve all members of the organization in evaluating organization wide results and in developing alternatives for making organization improvements. An action team can be appointed to consider the alternatives and make recommendations to top management regarding the high priority issues that need to be addressed in the change program.

Another action team should be appointed to evaluate Structural Issues such as the alignment between the mission, philosophy and goals and what the organization structure in fact values, the alignment of the structure with organization cultural values, and improvements that can be made to

Table 2. The Planned Change Approach

Leadership Excellence

1. The CEO and Top Management Team need to commit to taking a leadership role by planning activities that will provide the organization with *Vision, Direction,* and *Inspiration* and by *Championing* the pursuit of excellence.
2. The Top Management Team should define the Organization *Mission, Philosophy,* and *Goals,* reduce them to a few pages so they can be easily understood, communicate them throughout the organization, and continuously train employees to understand and implement them. Employees should have some level of involvement in establishing the Organization Mission, Philosophy, and Goals.

Structural Excellence

1. An Action Team should be formed to evaluate and monitor the organization structure and assure that it is (1) *designed* for results; (2) *aligned* to be consistent with and value behaviors implied or stated in the Mission, Philosophy, and Goals; and (3) *streamlined* to work effectively and efficiently and to be capable of responding quickly to needed change.
2. The Organization Philosophy should be incorporated into the Performance Review System so employees will become familiar with the philosophy and can be valued for supporting and practicing the philosophy. Continued noncompliance should be addressed. If there are no rewards for compliance or consequences for continued noncompliance, the philosophy will have little impact.

Team Excellence

1. Teamwork and Teambuilding should begin at the top. If top management is unwilling to commit to being a High Performance Team, it is difficult, if not impossible, to promote teamwork and cooperation throughout the organization.
2. Teamwork needs to be developed and encouraged both *within* and *among* groups. Teamwork should also be developed with clients, vendors, boards, unions, and other groups that could influence the success of the organization.

Management Development

1. A Management Philosophy should be developed and incorporated into the Performance Review System for managers and supervisors.
2. Managers and supervisors at all levels of the organization need to be trained in state-of-the-art thinking on management and leadership. Training should be selected that is consistent with and provides skills for implementing the Organization Mission, Philosophy, and Goals.
3. Managers and supervisors need to be encouraged to keep up-to-date on management. Training books, journals, tapes and films should be made available. Discussions, forums, and other opportunities for sharing management ideas should also be encouraged. A mix of internal and external training is desirable. Internal training provides continuity throughout management and external training exposes managers to additional ideas and perspectives.

Personal Excellence

1. Excellence programs are often management focused and may not reach all levels of the organization. Efforts need to be made to involve all levels of the organization in the excellence program and to promote Personal Excellence as well as Organization Excellence.
2. Employees at all levels should be involved in sharing ideas and in serving on Action Teams established to solve problems, make improvements, and innovate new ideas and should be included in training programs that promote excellence or could lead to improved performance and motivation.
3. Efforts should be made to seek ways to improve personal performance and development.

structure the organization for results (organization design issues, eliminating red tape, improving the reward systems, etc.). When structure reinforces desired changes, change occurs rapidly. The quickest and perhaps most effective vehicle for reinforcing change is the *Performance Review System.* Ideally, the Performance Review System should be redesigned to clearly reflect the values and desired behaviors implied in the organization philosophy. One simple way to accomplish this important step is to agree on desired *Behavioral Standards* that can be included as a one page attachment to the existing Performance Review System. Behavioral Standards clarify the behaviors expected of all employees at all levels in accomplishing their jobs that are consistent with the organization philosophy and cultural values. Examples are shown in Figure 2.

Teamwork is addressed next. The top management team is involved in team building sessions so they can become a role model for teams throughout the organization. Unless the top management team is willing to commit to teamwork, it would be hypocritical to expect others to place a high value on teamwork. Teamwork is then developed within and between teams throughout the organization and a methodology is created for managers and project leaders to use in quickly building temporary as well as continuing teams. A *team building methodology* can be developed by assigning a representative task group the responsibility of designing a handout for teams throughout the organization that includes: (1) *Guidelines For Team Leaders* on how to successfully lead teams; (2) *Team Standards* that describe team processes characteristic of high performance teams; and (3) *Guidelines For Structuring Teams for Results* that provide a format for establishing the purpose of the team, responsibilities of the team and team members, team goals, and other structural issues such as meeting times.

Examples of possible Team Standards are shown in Figure 2. Team standards establish positive norms for groups, bring out the best in team members, train them to be effective team players and provide a structured approach to periodically evaluating team processes.

The next two targets of change—*Management Excellence* and *Personal Excellence*—focus on individual change. Changes in management behavior can be accelerated by agreeing on *Management Standards* that describe what managers need to do to excel as managers in the organization. (see Figure2). The standards should be used to communicate to managers, supervisors and project leaders what is expected, and as a frame of reference for designing training and evaluating training alternatives. The Management Standards, in essence, represent a Management Philosophy that can be clearly communicated and that provides continuity and consistency throughout the organization regarding how management is practiced. Ideally, the Management Standards should also be integrated into the Performance Review System used for evaluating managers by including a one page attachment describing the standards.

IMPORTANCE OF STANDARDS

Standards can be used to communicate expectations, shape and reinforce cultural values, establish organizational norms, improve consistency throughout the organizations, and provide guidelines for how to succeed and improve performance. Standards should be designed to build champions and not compromised to value mediocrity.

A scale should be created (for example: 1 = Opportunity For Improvement; 2 = Achieves Expectations; 3 = Exceeds Expectations) so the standards can be used to evaluate performance. The Behavioral and Management Standards are most likely to be used when they are included as an attachment to the existing Performance Review System.

BEHAVIORAL STANDARDS

___ Committed to personal excellence and giving his/her best
___ Committed to being a team player.
___ Client centered and service oriented internally and externally.
___ Problem solver who is part of the solution and not part of the problem.
___ Good at encouraging and valuing others.

TEAM STANDARDS

___ The team has a clear purpose, clear responsibilities, and clear goals.
___ All team members participate and contribute to the success of the team.
___ Continued non-compliance with team standards is confronted and resolved.
___ Team members feel free to communicate openly, express their ideas, and to be creative.
___ The team is highly productive and successful at accomplishing goals.

MANAGEMENT STANDARDS

___ Provides leadership (vision, direction, and inspiration).
___ Assures that employees are clear on their responsibilities and goals.
___ Looks for the best in employees, tries to catch them doing things right, and makes efforts to develop their potential.
___ Recognizes and values high performance and good attitudes.
___ Quick to take care of performance or attitude problems and will not tolerate continued non-compliance.

Figure 2. Examples of Behavioral, Team, and Management Standards

Change programs often stagnate at the management level and never penetrate the remainder of the organization. The full benefits of change programs are only achieved when change results in *Personal Excellence* at all levels of the organization. This can be accomplished by involving employees throughout the organization in evaluating the strengths and opportunities for improvement of the organization and their own department and in exploring ideas for making needed improvements. It is also helpful to provide training that would result in personal and organizational improvement, to celebrate major successes or accomplishments and to expose employees to some of the fine videos available that address a wide variety of relevant issues such as organization excellence and service excellence.

Stage III—Transition

Once changes are made, it is often assumed they will last. However, unless changes are stabilized, they will not last! Therefore, the Transition Stage is essential to the change process. Stage III includes *Follow-Up* and *Evaluation.*

In the Follow-Up Phase, a strategy should be developed and implemented for championing, energizing, practicing, reinforcing, refining and culturizing the changes. It is particularly important that a person or team be appointed to champion follow-up activities. Without "Change Champions," change efforts rarely realize their potential.

In the Evaluation Phase, the changes and change program need to be evaluated. An evaluation, which often includes a second diagnosis, makes it possible to evaluate what changed and what remains to be changed and to learn from the change process. Perhaps of equal importance is the incentive and accountability a follow-up diagnosis provides for making needed changes if it is known early in the program it will occur.

The last step in the Evaluation Phase is to develop future plans for continuing to develop and improve the organization. A fatal mistake is to view Stage III as the end of a change program. Change management is an ongoing process of preparing, developing and improving an organization so it can function at or near its potential.

CONCLUSION

The highly competitive and volatile environment in which high tech firms operate places a premium on understanding how to manage change and develop high performance organizations. Top level managers need to be trained in change management and committed to being change champions, and the firm needs to have a sound strategy for continuously developing and improving the organization.

Upon contrasting my experiences in working with high tech firms versus a variety of other organizations, I have observed that the high tech firms appear to be forerunners of organizations of the future. Even more so, it appears their unique characteristics are increasingly becoming common to all organizations. If these observations are accurate, they may have significant implications. Perhaps the accelerated approaches to managing change that are evolving to respond to the needs of the high tech organization need to be more closely scrutinized and integrated into traditional thinking about the management of change.

REFERENCES

Bennis, W., & Nanus, B. (1985). *Leaders.* New York: Harper & Row.

Burke, W.W. (1987). *Organization development.* Reading, MA: Addison-Wesley.

Carlzon, J. (1987). *Moments of truth.* Cambridge, MA: Ballinger.

French, W.L., & Bell, C.H., Jr. (1984). *Organization development.* Englewood Cliffs, NJ: Prentice-Hall.

Huse, E.F., & Cummings, T.G. (1985). *Organization development and change.* St. Paul, MN: West Publishing.

Kanter, R.M. (1983). *The change masters.* New York: Simon & Schuster.

Kilmann, R.H. (1985). *Beyond the quick fix.* San Francisco, CA: Jossey-Bass.

Kouzes, J.M., & Posner, B.Z. (1987). *The leadership challenge.* San Francisco, CA: Jossey-Bass.

Naisbitt, J., & Aburdene, P. (1985). *Re-inventing the corporation.* New York: Warner Books.

New Management. (1988). *5*(3).

Peters, T. (1987). *Thriving on chaos.* New York: Knopf.

Peters, T., & Austin, N. (1985). *A passion for excellence.* New York: Random House.

Peters, T.J., & Waterman, R.H., Jr. (1984). *In search of excellence.* New York: Warner Books.

Tichy, N.M. (1983). *Managing strategic change.* New York: Wiley.

Tichy, N.M., & Devanna, M.A. (1986). *The transformational leader.* New York: John Wiley & Sons.

Warrick, D.D. (1984). *Managing organization change and development.* Chicago: Science Research Associates.

Weisbord, M.R. (1987). *Productive workplaces.* San Francisco, CA: Jossey-Bass.

SOME CHARACTERISTICS OF ORGANIZATIONAL DESIGNS IN NEW/HIGH TECHNOLOGY FIRMS

Harvey F. Kolodny

High Technology is subject to a variety of interpretations and definitions. For the purposes of this paper, the term will be used in the sense of the technology being both new to the particular firm and based on some form of microprocessor or computer-based technology. As such, *new technology* might be a more appropriate term than high technology. Any term used and any definition advanced will be problematic in some situations. For this paper, the issue of definition will be side-stepped and all references will be to *new/high technology*.

NEW/HIGH TECHNOLOGY AND ORGANIZATIONAL FORMS

One of the exciting aspects of organizational studies today is the variety that exists in the structures and processes used to organize the work place. The change is particularly noticeable in the area of structure. Earlier, stable environments restricted the set of organizational forms in general use to just a few (Miles & Snow, 1984). While there were always some innovative managers who developed their own creative forms (Maslow, 1965), most organizations were content to either emulate the majority or feel satisfied with the fit they had in place between a relatively predictable external environment and a relatively constant organizational form.

The latter form was almost always a functional one, based on specialization and the division of labor. If the organization grew sufficiently in size, then the structure might be decentralized into divisions which, in turn, almost always repeated the functional structure internally (Rumelt, 1974).

Now organizations face uncertain and complex environments (Lawrence & Lorsch, 1967). Technological change, global competition and a resulting increase in variety are important ingredients of that volatile externality (Grayson & O'Dell, 1988). Technological change is exploding at a rapid enough rate that firms in the same business can be found to be addressing similar markets with very different process technologies. Despite this, no clear superiority in the use of one technology over another results because the technology must be implemented through an organizational arrangement. No one, as yet, is able to state unequivocally what technological solutions should be combined with which organizational designs to provide the most effective solutions.

Flexibility has begun to replace efficiency as one of the key objectives of both the manufacturing and service sectors. This objective is achievable because current software based technology is so much easier to reprogram than the previous generation of equipment, which was generally based on hard-wired controllers. This article does not attempt to address the range of technological alternatives available to organizational designers. With microelectronic processors at the base of their controllers, a wide range of technological alternatives can be generated for most applications (Walton, 1983).

Nevertheless, it is important to point out that technological choices are not neutral (Patterson, 1983; Zuboff, 1985). They are loaded with social and, particularly for our interest here, organizational implications. It is this interdependency that makes the task to be undertaken here a bit of a fool's errand. *The task here is to identify organizational characteristics associated with new/high technology and indicate how organizational forms normally used by new/high technology organizations are more or less in support of these characteristics.*

The discussion that follows is problematic, for several reasons. If many of the new technologies in and of themselves determine some of the organizational forms, or in biological terms, if particular technological environments, within which specific technologies arise, select out specific organizational forms, then the two variables under study here—new/high technology (and its characteristics) and organizational forms—may be highly interdependent.

Two examples of how generic technologies might affect generic categories of organizational structures follow. The first example is products that are based on very rapid changes in their technology, for instance, navigation and control equipment. These product lines can force key lower-level decision makers (e.g., product designers) out onto the boundaries of the organization in order to stay in close contact with their technology sources (which would range from academic conferences to component suppliers). This behavior forces organizational structures to become more organic and less mechanistic (Burns & Stalker, 1961; Lawrence & Lorsch, 1967). The second example comes from the domain of process technologies designed to reduce cycle times, as in the case of just-in-time systems. These technologies force production organizations to become product-focussed (Kolodny, 1986a), to be structured "with the grain" (Lindholm, 1975), that is, with the direction of the flow of production, which of itself narrows the set of available organizational design choices.

Two more specific examples of technology determining the kinds of organizational forms discussed later in this article can be offered. Electronics high technology environments, such as the Boston area's "route 128" or the San Francisco area's "silicon valley," lead to a proliferation of "network" organizational forms, while the military and aerospace equipment sector tends to adopt "project management" organizational forms.

The other problematic aspect of this discussion is that technological determinism is indicated at a very time when such determinism is under solid challenge, both by technologists (Patterson, 1983) and, particularly, sociotechnical theorists (Davis, 1983-1984; Emery & Trist, 1960; Trist, 1981). Sociotechnical theorists maintain that technology should not be deterministic, that there is always organizational choice (Trist, Higgin, Murray, & Pollock, 1963), and that organizational effectiveness is better achieved through "joint optimization" of the social and technical subsystems than by designing around the technical variable alone.

The position advocated here is that technological environments will almost always dictate organizational forms, but more as "envelopes" of structural forms within which a variety of processes and approaches are possible. Sociotechnical theorists who put forward the argument of joint optimization do so more at a conceptual level than at an applied one, and would generally acknowledge that technological choices tend to "lead" social systems choices, even if they do not fully determine them. This line of discussion, however, won't be pursued here.[1]

With the awareness, then, that on the one hand, the technology of the new/high technology firms under study here has a somewhat deterministic effect on the organizational form alternatives to be considered, and on the other hand, that it is also a factor in developing the list of organizational characteristics to be evaluated (see below), I will nevertheless attempt to relate the organizational characteristics to the organizational form alternatives as if the two were relatively independent sets of variables.

ORGANIZATIONAL FORM ALTERNATIVES

The rationale for selecting the set of organizational forms that follow, as typical of new/high technology organizations, is a subjective one. However, it is drawn from personal experience with and research into the different forms (Armstrong & Kolodny, 1985; Kolodny, 1979, 1984, 1986a, 1986b; Kolodny & Dresner, 1986; Kolodny & Stjernberg, 1986; Liu, Denis, Kolodny, & Stymne, 1990).

The forms selected are listed in Table 1. They are in no particular order. Project management remains one of the most attractive ways to organize for high technology, as it has been since it first became a favorite of the military and aerospace industries (and, some would say, those who first constructed the Pyramids). Project management concepts, techniques and experience have been very well documented (e.g., Cleland & King, 1982). Matrix organization experience soon followed, with the form serving as a way to contain a variety of project, program and product undertakings (Kolodny, 1979) and with some arguments advanced about the form's particular relevance to successful new product implementation (Kolodny, 1980). Matrix organization research has also been well documented (Cleland & King, 1985; Davis & Lawrence, 1977; Knight, 1977).

Organismic is, for want of a better descriptive title, a way to refer to the organizational form of one successful auto parts manufacturer with one billion dollars in sales. In this company over one hundred relatively autonomous, small factories, each quite narrowly dedicated to a particular product or assembly and each with usually not more than 100 employees, have been bunched together into marketing groups that each have some particular product-market logic. Some functions are under corporate control (finance, human resources, legal), some are under group control (marketing, R&D), while the production function is under the almost complete control of each factory manager.[2] An elaborate incentive system for factory managers and a restriction on plant size has forced the organization to grow organically (at a current rate of about 15 or more factories per year). The form deserves to be in the set because of the extensive use of new technology to gain market share, in manufacturing processes, in product materials and particularly in the design function that uses computer aided design and manufacturing and computer aided process planning. There are a few recent descriptive articles of one particular company (e.g., Armstrong & Kolodny, 1985) and rumors of others who have adopted the form. Figure 1 is a schematic illustration of the particular company's "organismic" structure.

Network organizations have had considerable recent attention from researchers (Aldrich & Whetten, 1981; Miles & Snow, 1986; Whetten, 1981), but despite the research proliferation, they have remained difficult to conceptualize and classify. Networks are often the organizational mechanism

Table 1. Organizational Form Alternatives in New/High Technology Organizations

Organizational Form	*Area of Application*	*Size of Organization*	*Key Reason for Adopting Form*
Project Management	Professionals & specialist & production personnel	Small-Medium	Achieve integration
Matrix	Professionals & specialists	Medium-Large	Add variety
Organismic	Group & corporate level	Medium-Large	Grow horizontally
Network	Market & Professional level	Small	Gain leverage (on small size)
Product Focussed	Plant level and/or manufacturing level	Small	Reduce cycle time
Functional & Teams	Plant level	Small-medium	Achieve coordination

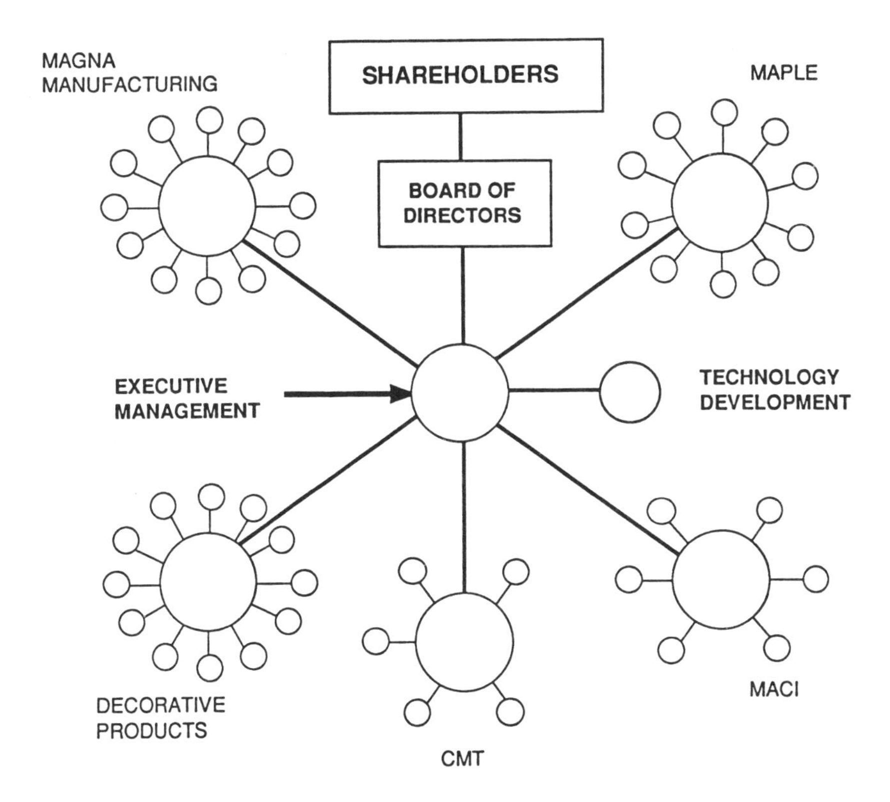

Source: Adapted from *Annual Report,* 1983.

Figure 1. Magna International Inc.'s "Organismic" Structure

used by small high technology firms to bid on quite large contracts, which they achieve by "networking" or joining-up with other similarly oriented firms. In a subsequent bid, the combination of players may differ, though networks tend to maintain the same partners when trust and reliability have been established. Network organizational forms have also served as a way to classify "hollow" corporations (*"The Hollow Corporation,"* 1986), to conceptualize technology transfer organizations, and they are probably a useful concept to contain the increasingly popular notion of "strategic alliances."

Product focused forms include such subclassifications as semi-autonomous work groups, group technology cells, parallelization and product shops (Aguren & Edgren, 1980; Kolodny, 1986a, 1986b). They are structural forms "organized with the grain," meaning, designed to follow the flow of production. One of their primary attributes is their speed of response. Cycle time reduction factors of six and seven and even ten frequently result from reorganization to a product focussed form, though this is usually accompanied by the introduction of new process manufacturing technology such as computerized numerical control machine tools with automatic tool changers or automated printed circuit insertion equipment. The restructuring is also usually accompanied by materials and inventory management changes and even comprehensive computer integrated manufacturing (CIM).

Functional forms with teams refers to organizational structures recently implemented in continuous process plants where, more often than not, the process technology is controlled by a digital computer. Operator roles, and even some maintenance roles, have altered to where the task is now one of observing schematics of the process or readouts of functions on visual display terminals (Zuboff, 1988). High capital costs for plant and equipment have encouraged firms to assign more responsibility to well trained operators, and desires for improved quality of working life have also contributed to the decentralization of more responsibility to the operator level; but the complexity of the production process and the unpredictability of external inputs have made team operation a logical solution to the problem of having the total knowledge of the system divided among several operators (Davis, 1983-84). The responses required are holistic ones, or systemic ones, and team organization facilitates this outcome. Furthermore, though the hierarchies of these new plants are relatively flat, the team structure still supports a faster response than even the flattened functional hierarchy could provide.

This latter category, functional forms with teams, appears often in the many new plants that are called high commitment work systems (Walton, 1985) or high-involvement management (Lawler, 1986) or sociotechnical systems designs (Cherns, 1976; Davis, 1983). However, it is the organizational form that is being addressed here, not the process of designing, developing and sustaining it. Since high commitment, high involvement and sociotechnical systems are processes associated with the design and maintenance of work systems (Taylor & Asadorian, 1985), rather than particular structural arrangements, the form is referred to as "functional with teams."

The three other columns in Table 1 help to differentiate the forms from one another. There are some caveats to the comments in each column. Under "Area of Application," the suggestions are meant to be typical. The forms may have application in areas of the organization other than those listed.[3] In the second column, the weightings assigned to the variable "size" are very subjective. The particular attribution made is based on a mixture of items such as: number of employees, capital investment, relative plant size and a perspective on the unit under study with respect to the rest of the organization when that is the appropriate context. Particular forms are usually selected out for more than the single reason shown in the third column,[4] but for purposes of parsimony one strong reason is cited in the last column of Table 1 for adopting the particular form. It is often the dominant reason.

ORGANIZATIONAL CHARACTERISTICS OF NEW/HIGH TECHNOLOGY FIRMS

Are the organizational characteristics of firms in new/high technology areas different than those using older or conventional technology? Since the dependent variable in this question is "organizational characteristics," the answer is probably "no." The characteristics that are basic to organizations are likely constant.

Nevertheless, new/high technology is bound to have some impact on the character of an organization. Boddy and Buchanan (1986, pp. 24-26), referring specifically to information technology, suggest new technology has "enabling characteristics" rather than determining features. For example, it can facilitate convergence (of telecommunications and computing); it can facilitate integration (of discrete stages of manufacturing or administrative processes); and it has high innovative potential. In reviewing the many recent articles on new/high technology organizations (Gomez-Mejia & Lawless, 1988), a variety of terms are repeatedly referred to as characteristics of these firms, among them: flexibility, responsiveness, adaptiveness and complexity management.

Several of these cited characteristics have considerable similarity in their meaning. Some overlap each other. Some subsume others. After examining a range of the characteristics frequently cited as associated with new and/or high technology, a subjective grouping was attempted (a kind of intuitive and manual factor analysis) and four reasonably independent variables emerged. The variables are listed in Table 2 with the more descriptive words that characterize them, and to some extent, bound them.[5]

Two of the variables listed should not be surprising to organizational theorists. Specialization and coordination or differentiation and integration are basic to all organizational activity once the organization gets beyond a size where these can be contained in the minds and actions of one or just a few people. However, it is worth remarking on the words used to describe these two variables. Variable B recognizes the market responsiveness of organizations in highly differentiated and highly changing environments. It also includes the increasing complexity of new/high technology organizations, particularly the need to manage complementary orientations simultaneously. This is a significant divergence from traditional, hierarchical structures. Variable C, which includes networking, extends the subject of coordination beyond the firm's borders, again a divergence from traditional organizational thinking.

The temporal orientation of variable A is less attended to in organizational studies than the two variables described above. However, flexibility and speed of response, brought about by rapid technological change, have come to characterize new/high technology firms. They are increasingly important attributes.

Variable D will be even less familiar to many. It has received more attention of late because of the increasing appreciation of how little of the potential of people in organizations is being realized. It also delves deeper into relationships, both with the organization's internal members and its external ones. The words that describe the variable are still "immature." The increasing number of high involvement, high commitment or sociotechnical systems designs indicate that this variable should receive extensive study in the near future.

ORGANIZATIONAL FORMS AND ORGANIZATIONAL CHARACTERISTICS

Table 3 relates alternative organizational forms to the different organizational characteristics. For the scale used, a "5" means that the organization form is high in its support of the particular variable. Hence, project management is viewed as an organizational form that highly supports time-related characteristics (Variable A) such as flexibility, responsiveness, and so forth. Or in another example, organismic organization forms are viewed as only moderate in their support of differentiation characteristics (Variable B).

The assessments are the author's subjective interpretations. If all four organizational characteristics were equally important, a sum of the scores

Table 2. Organizational Characteristics of New/High Technology Firms

Variable	*Characteristics*
Variable A Time Related	• flexibility • responsiveness (to customers, competitors, technological change) • risk-taking behavior • mobility, impermanence, adaptability • entrepreneurial behavior
Variable B Differentiation	• variety increasing properties (in products, in processes, in markets) • management of complexity • management of complementarities (stability and flexibility, open processes and standardization, product, function and area choices, social and technical) • information processing capacity (uncertainty, variety) • diversification, decentralization • experimentation • market driven
Variable C Integration	• integration approaches • networking • communication and coordination • decision-making
Variable D Investment in Organization	• investment in "organization" • development of internal relationships • development of external relationships (to clients, suppliers, technology sources) • control systems • interpersonal processes • local knowledge

Table 3. Organizational Forms and Organizational Characteristics

	Organizational Characteristics			
Organizational Forms	*Variable A (time-related)*	*Variable B (differentiation)*	*Variable C (integration)*	*Variable D (investment in organization*
Project management	5	4	4	3
Matrix	3	5	4	5
Organismic	5	3	2	4
Network	3	2	5	3
Product focussed	5	1	5	3
Functional & teams	3	1	4	3

Notes: 5 = High; 4 = Moderate-High; 3 = Moderate; 2 = Low-Moderate; 1 = Low.

would indicate how effective one organization form is with respect to the other forms in supporting the characteristics of new/high technology organizations. However, the variables are not of equal importance. Furthermore, new and high technology have been aggregated in this paper and if they were separated, the assigned weightings might differ (so, too, might the list of characteristics).

CONCLUSIONS

A variety of organizational forms are available to organizational designers and some types are selected out quite consistently in situations of new and of high technology. They are probably selected, after trial and error attempts at other organizational arrangements, because they best provide the structural arrangements and the processes that the particular organization needs to accomplish its tasks. This paper identifies and describes some of those forms. The paper also identifies key organizational variables frequently associated with new/high technology firms. The particular groupings are subjective assessments in need of testing. The paper then goes on to rate the forms in terms of how well they support the variables. The basis of the assessment is subjective, but it draws extensively on the author's experience with new/high technology situations. It would be difficult to test for the weightings assigned, if only because of the difficulty in agreeing on effectiveness criteria. However, the table of relationships (Table 3) should make organizational designers think harder about their selection or organizational forms.

This is a world in which there are now many more choices about organizational alternatives (forms) than there have been in the recent past. It is also a world in which technological alternatives are many and the variations are proliferating. It is probably going to require some form of "organization design skill" to achieve a good fit between the organizational and technical alternatives available. Schon (1983) refers to these skills of "managing complexity" as "reflection-in-action." To begin to acquire these skills, designers will have to have thought about the relationships discussed in this paper.

ACKNOWLEDGMENT

An earlier version of this paper was presented at the Managing the High Technology conference in Boulder, Colorado (June 13-15, 1988) and published in *Proceedings: Managing the High Technology Firm Conference* (see Gomez-Mejia & Lawless, 1988).

NOTES

1. In fact, with the flexibility that now exists in microprocessor-based software controlled technology, designers who are sensitive to social system concerns can continue to lead via the

technology and still arrive at very satisfactory sociotechnical solutions (Walton, 1983). As a result of this technological flexibility, the somewhat idealistic objective of joint optimization has become considerably more realistic.

2. Magna International is an excellent illustration of "markets and hierarchies" organizational theory (Williamson, 1975). Although plants are autonomous, human resource activities are tightly controlled from the corporate office. On the other hand, plants within the same group compete ferociously with each other for business, as if in the public marketplace.

3. One assembly factory, whose organizational form can be classified as both "product focussed" and "functional with teams", has a "matrix" design that operates down to the blue collar level, which explains why it can be simultaneously functional and product organized (Kolodny & Dresner, 1986).

4. See Lawrence, Kolodny and Davis (1977) for the multiple reasons for adopting a matrix organization design and Kolodny (1979, Table 1) for the multiple determinants of functional and project forms.

5. Risk-taking and entrepreneurial behaviors are included in Variable A for parsimony, though one might argue that they better belong with yet another independent variable, one that would include financing aspects and risk assessments of the new/high technology.

REFERENCES

Aguren, S., & Edgren, J. (1980). *New factories*. Stockholm: Swedish Employers' Confederation.

Aldrich, H., & Whetten, D.A. (1981). Organization-sets, action-sets, & networks: Making the most of simplicity. In P.C. Nystrom and W.H. Starbuck (Eds.), *Handbook of organizational design* (Vol. 1, pp. 385-408). Oxford: Oxford University Press.

Armstrong, A., & Kolodny, H. (1985). *Magna International Inc.* Case study, Faculty of Management, University of Toronto.

Boddy, D., & Buchanan, D.A. (1986). *Managing new technology*. Oxford: Basil Blackwell.

Burns, T., & Stalker, G. (1961). *The management of innovation*. London: Tavistock.

Cherns, A.B. (1976). The Principles of Sociotechnical Design. *Human Relations, 29*(8), 783-792.

Cleland, D.I., & King, W.R. (1982). *Project management handbook*. New York: Van Nostrand Reinhold.

_______(1985). *Matrix management handbook*. New York: Van Nostrand Reinhold.

Davis, L.E. (1983). Learnings from the design of new organizations. In H.F. Kolodny & H. Van Beinum (Eds.), *The quality of working life and the 1980's*, (pp. 65-86). New York: Praeger.

_______(1983-1984). Workers and technology: The necessary joint basis for organizational effectiveness. *National Productivity Review, 3*(1), 7-14.

Davis, S.M. & Lawrence, P.R. (1977). *Matrix*. Reading, MA: Addison-Wesley.

Emery, F.E., & Trist, E.L. (1960). Socio-technical systems. In C.W. Churchman and M. Verhurst (Eds.), *Management science models and techniques*, (pp. 83-97). London: Pergamon Press.

Gomez-Mejia, L.R., & Lawless, M.W. (Eds.). (1988). *Proceedings: Managing the high technology firm conference*. Boulder, Co: University of Colorado Press.

Grayson, C.J. Jr., & O'Dell, C. (1988). *American business: A two-minute warning*. New York: The Free Press.

The hollow corporation: Special report. (1986, March 3). *Business Week*, pp. 57-66, 71-74, 78-85.

Knight, K. (Ed.). (1977). *Matrix management: A cross-functional approach to organization*. Farborough, UK: Gower Press.

Kolodny, H.F. (1979). Evolution to a matrix organization. *The Academy of Management Review, 4*(4), 543-553.

———(1980). Matrix organization designs and new product innovation success. *Research Management, 23*(5), 29-33.
———(1984). *Product organization structures improve the quality of working life* (Occasional paper No. MM84-636). Dearborn, MI: Society of Manufacturing Engineers.
———(1986a). Product focussed forms. In *Flexible manufacturing cells '86: Proceedings of the computer and automated systems association of the society of manufacturing engineers* (pp. 1-19). Dearborn, MI: Science Manufacturing Engineers.
———(1986b). Assembly cells and parallelization: Two Swedish cases. In O. Brown, Jr. & H.W. Hendrick (Eds.), *Human factors in organizational design and management - II* (pp. 521-526). Holland: Elsevier.
Kolodny, H.F., & Dresner, B. (1986). Linking arrangements and new work designs. *Organizational Dynamics, 14*(3), 33-51.
Kolodny, H.F., & Stjernberg, T. (1986). The change process in innovative work design in Sweden, Canada and the USA. *The Journal of Applied Behavioral Science, 22*(3), 287-301.
Lawler, E.E., III. (1986). *High involvement management.* San Francisco, CA: Jossey-Bass.
Lawrence, P.R., Kolodny, H.F., & Davis, S.M. (1977). The human side of the matrix. *Organizational Dynamics, 6*(1), 43-61.
Lawrence, P.R., & Lorsch, J.W. (1967). *Organization and environment.* Cambridge, MA: Harvard Graduate School of Business Administration.
Liu, M., Denis, H., Kolodny, H., & Stymne, B. (1989). Organization design for technological change. *Human Relations, 43*(1), 7-22.
Lindholm, R. (1975). *Job reform in Sweden.* Stockholm: Swedish Employers Confederation.
Maslow, A.H. (1965). *Eupsychian management: A journal.* Homewood, IL: Richard D. Irwin.
Miles, R., & Snow, C. (1984). Designing strategic human resources systems. *Organizational Dynamics, 13*(1), 36-52.
———(1986). Organizations: New concepts for new forms. *California Management Review, 28*(3), 62-73.
Patterson, W.P. (1983, May 30). Where is technology taking us? *Industry Week,* pp. 30-40.
Rumelt, R.P. (1974). *Strategy, structure and economic performance.* Cambridge, MA: Harvard Graduate School of Business Administration.
Schon, D.A. (1983). *The reflective practitioner.* New York: Basic Books.
Taylor, J.C., and Asadorian, R.A. (1985). The implementation of excellence: STS management. *Industrial Management, 27*(4), 5-15.
Trist, E.L. (1981). *The evolution of socio-technical systems.* Toronto: Ontario Quality of Working Life Centre.
Trist, E.L., Higgin, G.W., Murray, H., & Pollock, A.B. (1963). *Organizational choice.* London: Tavistock.
Walton, R.E. (1983). Social choice in the development of advanced information technology. In H.F. Kolodny & H. Van Beinum (Eds.), *Quality of working life in the 1980's* (pp. 55-64). New York: Praeger.
———(1985). From control to commitment in the work place. *Harvard Business Review, 63*(2), 77-84.
Whetten, D.A. (1981). Interorganizational relations: A review of the field. *Journal of Higher Education, 542*(1), 1-28.
Williamson, O.E. (1975). *Markets and hierarchies: Analysis and antitrust implications.* New York: The Free Press.
Zuboff, S. (1985). Automate/informate: The two faces of intelligent technology. *Organizational Dynamics, 14*(2), 5-18.
———(1988). *In the age of the smart machine: The future of work and power.* New York: Basic Books.

PART IV

HUMAN RESOURCE MANAGEMENT IN HIGH TECHNOLOGY FIRMS

STRATEGIC HUMAN RESOURCE MANAGEMENT IN HIGH TECHNOLOGY INDUSTRY

Wayne F. Cascio

A popular mystique has developed around the creative geniuses who make the initial technical breakthroughs and the entrepreneurs who make and market the products. Applications of these innovations on a wide scale—their design, production, and distribution—depends, in the final analysis, on large numbers of people within organizations, performing interrelated functions. It is people who sustain new firms, create new industries and new markets, and shape the overall impact of these innovations.

— Kleingartner and Anderson (1987, p. viii)

If people are so crucial to the success of high technology firms, then certainly human resource management (HRM) strategies for attracting, retaining and motivating people also are critical to gaining and sustaining competitive advantage. Who are these people who work in high technology industry, and why are they so valuable? According to Kleingartner and Anderson (1987) we may think of high technology industries as those that share these attributes:

- The proportion of engineers, scientists and technicians in their work force is higher than in other manufacturing industries.
- They are science-based in the sense that their new products and production methods are based on applications of science.

- Research and development are more important to their successful operation than they are to other manufacturing companies.
- They depend on the academic community to educate their work force to a larger degree than traditional manufacturing companies do.
- The markets for their products are both national and international.
- The life of their products tends to be short, with products often becoming obsolete before mass production can be undertaken.

In sum, high technology firms are driven by three things: innovation, science and research. Given these special characteristics, it is not surprising that HR professionals working in such firms consistently stress three high-priority issues (Miljus & Smith, 1987):

1. *Recruitment and staffing*—the critical need to attract and properly place high-talent employees or "knowledge workers" throughout the organization.
2. *Training and development*—the requirement for firms continually to upgrade, broaden and deepen the technical skills of knowledge workers.
3. *Organization design and development*—the need to work closely with line managers to shape and to maintain organizational conditions that support innovation, change and employees' continued high performance.

These three issues suggest a rich agenda of strategic decisions for HRM. First we will consider some key challenges in each of these areas, and then we will show how they are being addressed in one high technology company, Hewlett-Packard Singapore.

RECRUITMENT AND STAFFING

Competition for knowledge workers is intense, not only among "mainline" high technology firms, but also among academic research centers, manufacturers of automobiles, machine tools and equipment, aerospace equipment and specialty steel. Here are four key areas in which labor markets are particularly "tight."

1. *Manufacturing engineering.* This is a relatively new specialization that involves designing a product with an understanding of how it would be manufactured. Although several leading schools are developing degree programs in the field, the demand for graduates far outstrips the supply of them.
2. *Plastics.* Although there are several schools with graduate programs in polymer science, the basis of the plastics industry, almost all such programs are undersubscribed, despite keen competition for graduates.

3. *Electrical engineering.* According to the *Wall Street Journal*, the serious shortage in this area is expected to get worse because demand for electrical engineers is increasing—not only in high technology fields, but also in conventional manufacturing. At the same time, severe faculty shortages in graduate schools (see below) limits the number of candidates being turned out (Morgenthaler, 1986).
4. *Cutting-edge technologies.* Specialists are needed in optics, laser technology, electromagnetics, avionics, and composites technology. Composites technology involves the development and application to manufacturing of new materials. An example is the use of composites in aircraft structures that are lighter and stronger than any material used in the past (Morgenthaler, 1986).

Demand for trained talent in other areas is almost as great. For example, consider that over the five-year period from 1982 to 1987, there was a compound increase of over 100 percent in the demand for hardware and software engineers in high technology firms. At an increasing rate, much of this talent is foreign, educated in the United States, and predominantly Asian, with no dilution in quality. These individuals represent the top 1 percent of students in their countries; they are the "cream of the crop." In engineering, for example, an extraordinary 55.4% of last year's doctorates went to candidates from overseas (at Penn State, the figure was 74%). By 1992, Iowa State University predicts that somewhere between 75 and 93% of engineering professors will be foreign born. Not only are universities running out of homegrown talent, but recruiters for some of the country's leading high technology firms say they are unable to find a single qualified American to hire (*"Wanted,"* 1988).

Scenarios like these suggest that the marketplace for recruiting high talent human resources is international, even global, in scope. A variety of innovative inducements have been offered to high tech recruits, even before they graduate. For example, some firms have provided company-sponsored scholarships and internships or cooperative arrangements for students to work on the company's challenging products. This suggests a "tighter coupling" between high technology firms and the universities that produce the human talent to make them go. This coupling can assume several, interrelated forms, including equipment grants and monetary support for faculty research. All of this is oriented toward a single strategic objective: to gain and to sustain a competitive advantage by "locking-in" a regular flow of highly skilled, technically competent, human resources.

After graduation, technically educated candidates can expect to receive offers that include, for example, desirable geographic locations that afford favorable weather, cultural and ethnic diversity, accessible recreational alternatives and access to major research universities that provide exceptional academic opportunities for advanced studies and retraining. Companies that offer "cutting edge" technology and research opportunities, and that have established reputations for innovation and solid financial growth, typically enjoy an advantage over the competition.

Temporary and Part-time Employees

Thus far we have been focusing on the recruitment of professional and technical personnel. However, the industry also makes extensive use of temporary and part-time employees (Gomez-Mejia & Welbourne, in press). These jobs help to cushion firms against cyclical events in the economy and in the industry, as corporations strive to keep their full-time staffs as lean as possible. Temporary, part-time and contract workers now comprise nearly one-third of the U.S. work force. The most common reasons why companies hire them are: to alleviate an overload of work; for special projects; to cover for workers who are on leave; to cover for vacationing workers; to fill in for departing workers; to fill in for sick employees; and to perform duties where permanent jobs cannot be justified financially (McCarthy, 1988).

Yet firms report special problems with these workers. Quality of work and motivation are the two most formidable ones. The fastest growing segment of the pool of "contingent" workers is professional and technical employees—employees who often require firm-specific training and lengthy acclimation periods in order to be able to make meaningful contributions. Consider the dilemma faced by the computer systems development division of Federal Express Corporation. The division turned to temporary workers for several projects.

First, the temporary workers spent weeks learning the Federal Express computer systems. Then they designed new systems and left, leaving only a few supervisors who understood how things worked. The supervisors then spent weeks training the regular employees. Federal Express felt it had to break that cycle.

The company's solution was expensive, but it kept much of the flexibility of the contract workers: it started an in-house pool of part-time programmers. The company offered them benefits nearly identical to those of full-time workers, including medical and dental insurance and tuition refunds. This is a benefits package that is far more generous than most companies offer to part-timers. However, Federal Express can now tap a ready pool of experienced part-timers, all of whom who are eligible to fill full-time positions.

More on Recruitment

There are two additional recruitment issues that are critical to prospective candidates: the design of their jobs and their compensation for performing them. Realistic previews of jobs that are challenging, that provide autonomy in the context of team collaboration, real responsibility and control over some meaningful aspect of the work, are essential components of high technology recruitment. The perceptions of recruits regarding these issues are crucial determinants of the success or failure of recruitment efforts, as a recent study (Taylor & Bergmann, 1987) showed. *Job attributes* such as the type of supervision, the degree of challenge in the job, geographic location, salary and job title were much more important to job applicants than were recruitment *activities* such as the demographic characteristics of the recruiter, or the recruiter's behavior during job interviews. In short, the recruitment message predominates over its media.

COMPENSATION

Innovative compensation and benefits packages are required to attract and to retain highly talented engineers and scientists. Contrary to what might be expected, evidence indicates that high technology firms tend to adopt compensation strategies for engineers and scientists that are right at the 50th percentile of the market or slightly above it. They tend *not* to be market leaders in base salaries. Instead, since a primary objective of the compensation system is to promote and to encourage innovation, high technology firms use financial incentives, such as stock options and profit-sharing plans to augment base salaries and to create an attractive *total* compensation package for their professional personnel (Balkin & Gomez-Mejia, 1985).

In terms of benefits and perquisites, the Hay Group conducted a study that matched managers at high technology concerns against officials of all industrial companies. They found the high technology managers to be a breed apart. They were less likely to get chauffeurs, access to company planes or country-club memberships. However, they were more likely to be offered low-cost loans, additional medical insurance and extra schooling (*"High Tech,"* 1986).

The tendency to emphasize incentives over base salary is even more pronounced among small firms at the growth stage of their life cycles (Balkin & Gomez-Mejia, 1987). These firms are characterized by a high proportion of expenditures on research and development, high mortality rates and a dependence on product innovation in order to get established in their industries. However, even after controlling for sales volume, product life cycle, profitability and attrition rates, high technology firms place greater emphasis on incentive programs than do firms outside this industrial sector (Balkin & Gomez-Mejia, 1984). Finally, in addition to high base salaries and attractive incentives, high technology recruits can expect to receive such other inducements as sign-on bonuses, front-end paid vacations, settling-in allowances at the time of hire, long-term equity arrangements and short-term bonuses tied to individual and group performance (Miljus & Smith, 1987).

In order to optimize overall production costs, high technology firms often adopt similar strategies: retain research and development in the United States and locate standardized, high-volume production offshore. Buffa (1984) estimates that 85% of the 36 million people who enter the annual worldwide labor force are from Third World nations, where high-volume products are manufactured at costs significantly lower than U.S. production costs. To appreciate this, consider that while U.S. manufacturing workers averaged $9.85 per hour in wage costs alone in 1986, average hourly wage costs in other representative countries were as follows: Singapore: $2.23; Hong Kong: $1.89; Taiwan: $1.67; Korea: $1.39; and Thailand: $0.31 (National Trades, 1987).

More on Profit-sharing

One of the most intriguing questions surrounding the compensation practices of high technology firms is the effect of profit-sharing at lower organizational levels. However, recent data have begun to reveal some striking findings, as we shall see. To provide perspective on this issue, though, let us first define our terms.

Under a profit-sharing plan, employees normally receive a bonus that is based on some percentage (e.g., 10 to 30%) of a company's profits beyond some minimum level. Profit shares may be paid directly to employees at the end of the fiscal year (as is done by about 40% of all plans), but more often they are deferred. That is, they are put into a fund for retirement or death benefits. Proponents of profit-sharing cite two benefits:

1. It strengthens employees' sense of involvement with the enterprise, cuts waste and motivates employees to work harder.
2. The firm can provide pensions and other benefits without incurring fixed costs since contributions are made only in profitable years.

On the other hand, critics of profit-sharing point to four disadvantages:

1. The relationship between performance and reward is weaker than even group incentives provide.
2. There is a long delay between effort and reward since employees do not receive their share of company profits until a year after they earn them. This need not be the case, of course, as we shall see shortly.
3. Many employees do not understand how profits are computed. Further, if labor-management relations are poor and untrustworthy, then employees may suspect that profits are being underreported.

4. From the perspective of employees, benefits and pensions are insecure under deferred profit-sharing because the company pays only if it makes a profit.

Many companies have long had profit-sharing plans to supplement pensions. However, these plans tended to be reserved for top executives. Now, under the leadership of firms such as Hewlett-Packard, Ford Motor Company and American Telephone & Telegraph, almost everyone employed by these firms receives a bonus each year that profitability goals are met. Among high technology firms, Balkin and Gomez-Mejia (1984) found that 55% used profit-sharing for research and development employees, compared to only 33 percent in traditional firms. A more recent survey by Gomez-Mejia and Balkin (1989) found that 68% of scientists and engineers in high technology firms receive profit-sharing benefits in their compensation packages.

Many of the firms that are now using company-wide profit-sharing have undergone restructuring, downsizing, mergers and layoffs during the 1980s. They hope that profit-sharing will rebuild teamwork among surviving workers and get them to focus on how their performance relates to company profitability. They also recognize, however, that such plans will help them to control labor costs in the event of a downturn in business.

Profit-sharing makes total compensation costs more variable, especially when it is combined with merit-raise systems that replace automatic annual increases. This helps slow the growth of base salaries, as well as the cost of benefits that are tied to base salaries, such as life insurance.

Employees do not seem to be complaining, as data recently reported in *Business Week* (Schroeder, 1988) illustrate. For example, Ford Motor Company's recovery has rewarded all 156,500 eligible U.S. salaried and hourly workers for the past five years. In 1987 Ford disbursed checks that equaled 11% of paychecks—or an average of $3,700 per employee. Top labor relations executives believe that there is a link between Ford's profit-sharing and its 40% improvement in its break-even point for its North American automotive operations.

Empirical data support this conclusion. A study done at Rutgers University concluded that companies with cash profit-sharing had 10% higher increases in productivity than their competitors in the same industries who did not have plans. They also had one-third fewer layoffs in bad times because their labor costs shrank automatically (Schroeder, 1988).

At Hewlett-Packard, profit-sharing has proven to be a powerful incentive for workers to watch the bottom line instead of the clock. Said the firm's associate general counsel, "It's not unusual for someone to observe wasted stationery or a computer terminal that has been left on and complain, 'there goes our profit-sharing.'" Among research and development personnel, profit-sharing has been especially beneficial. These individuals tend to focus their energies on their current research projects and developments in their own scientific disciplines (Balkin, 1987). Profit-sharing forces them to look beyond the goals of their specific work group and to consider the implications of what they do on the total organization and also on the "bottom line." Profit-sharing plans may also provide a mechanism for enhancing cooperation and teamwork between work groups or departments. Social psychologists call this a "superordinate goal," for it is a broad objective with which all employees can identify. Hewlett-Packard has been paying bonuses for 25 years—with steady, positive gains in productivity. Other companies are now recognizing that it's good business to do the same for their employees.

TRAINING AND DEVELOPMENT

Training consists of planned programs designed to improve performance at the individual, group or organizational levels. Improved performance, in turn, implies that there have been measurable changes in knowledge, skills, attitudes or social behavior. Training and development is one strategy for improving the person/job/organization fit, but it is by no means the only one. In order to think strategically about training and development, firms must examine their needs at three levels: (1) at the level of the organization, it is important to identify *if* training is needed, and if it is, exactly where it is needed; (2) at the level of operations, it is necessary to identify the content of training; finally, (3) at the level of the individual, it is necessary to identify who needs training, and what kind. Each of these levels contributes something, but to be most fruitful all three must be conducted in a continuing, ongoing manner, and at all three levels: at the organization level—by top managers who set its goals; at the operations level—by managers who specify how the organization's goals are going to be achieved; and at the individual level—by managers and workers who do the work and achieve those goals (Cascio, 1987).

Many firms tend to view training as organizational firefighting. That is, they provide it in response to specific problems that surface or new technologies that are developed elsewhere. To remain viable and competitive, these firms recognize that they have to train. Consider just one example of this. As a result of continued development in office automation, employers will have to retrain office workers five to eight times during their careers (Wexley, 1984). While some such training clearly is necessary, if training is done *solely* as a response to actual or potential problems, it is reactive.

In the high technology industry, competitive pressures require organizations to be proactive—to anticipate training needs. Xerox has been doing this for years with its scientists and engineers by analyzing personal history items and demographic characteristics contained in its personnel inventory. Items such as continuing technical education programs attended, age and years since last degree are evaluated to identify individuals whose knowledge and skills potentially are obsolete. It then tailors training programs to the specific needs of these individuals (Morano, 1973).

Another change in the strategic thinking of high technology firms is to view training, development and personal growth in the context of long-term career planning. Such programs often are integrated with performance appraisals in which employees and their managers jointly examine the employee's strengths, future career options, relevant development needs and alternative strategies for providing them. High technology firms that are able to attract, retain and continually improve the skills of their knowledge workers tend to be successful in the long-run. High technology firms recognize this, and as a result, training and development receive high priority in them.

Heavy emphasis on training and development is not unique to high technology firms. Companies in all sectors of the economy are coming to regard training expenses as no less a part of their capital costs than plant and equipment. Total training outlays by U.S. firms are now $30 billion a year—and rising (*Serving the New Corporation,* 1986). Additional support for this conclusion comes from a recent survey of 300 Midwestern manufacturers. The top priority for spending over the next five years, ranking ahead of process simplification (such as just-in-time supply) and automation, is spending for training and development (*"Manufacturing Spending,"* 1988). The majority of the training will go to workers rather than supervisors. This is a significant change. In 1983, most of the training went to supervisors; in 1988, 65 percent went to workers. Why the increase? The complexity of new technology has simply overwhelmed workers. Managers now recognize that unless the workers are trained to perform at least to a minimum a level of competence in using the new technology, then most of its benefits will not be realized (*"Manufacturing Spending,"* 1988).

At the level of the individual firm, Motorola is typical. It budgets about 1 percent of annual sales (2.6% of payroll) for training. It even trains workers for its key suppliers, many of them small- to medium-sized firms without the resources to train their own people in such advanced specialties as computer-aided design and defect control. Taking into account training expenses, wages and benefits, the total cost amounts to about $90 million per year. Results have been dramatic, according to a company spokesperson: "We've documented the savings from the statistical process control methods and problem-solving methods we've trained our people in. We're running a rate of return of about 30 times the dollars invested—which is why we've gotten pretty good support from senior management" (Brody, 1987).

Training as an Antidote to Obsolescence

Knowledge workers drive high technology firms. Knowledge workers create the products, the processes and new technologies that change the industry. Coping with change—managing it and ultimately controlling it, is an ongoing challenge for high technology firms. Training is an antidote to the Paul Principle, which holds that over time people become uneducated, and therefore incompetent, to perform at a level that they once performed at adequately. That time frame is narrowing considerably. Because of the exponential rate of change in the electronic instrumentation and computation industry, knowledge workers become technically obsolete within three years following graduation from a baccalaureate engineering program (Miljus & Smith, 1987). Hence scientists and engineers must be given opportunities to upgrade their skills in areas such as systems architecture, design methodologies, operating systems, programming languages, and so forth.

Thus far we have been concentrating on avoiding *technical* obsolescence. But what about *managerial* obsolescence? Certainly the people-related skills necessary to manage the innovation process and new production processes are no less important. At New United Motor Mfg., Inc. (NUMMI), the Toyota-GM joint venture in Fremont, California, both managers and workers were retrained in Japanese-style team production methods. Harried foremen were no longer pushed to move as much metal as possible down the line and let inspectors at the end worry about defects. Instead, they were retrained as "group leaders" responsible for maintaining quality standards in the parts their section receives and the work it does. Selected workers who had little responsibility in the past have been trained as team leaders who head small groups of workers responsible for producing defect-free work in a series of assembly steps. The goal: to raise productivity and to reduce defects to Japanese standards. NUMMI is now approaching that after two years of production (Brody, 1987).

A wide range of new (and expensive) technologies is being used to present training—interactive video, computer-assisted instruction, teleconferencing—yet we know very little about the relative effectiveness of these techniques. Ongoing programs of evaluation research are needed to assess learning in training and the transfer of the new knowledge, skills or abilities back to the job.

Short product life cycles (three years or less for many new products in the electronics industry) complicate this process. To raise the knowledge worker's awareness of new technologies, market demands and process improvement methods, requires a changing cadre of trainers. This, in turn, suggests the need for a dedicated, organized approach to internal technical and managerial training in order to maintain and upgrade the caliber of the high technology work force.

UNION REPRESENTATION AND THE HIGH TECHNOLOGY FIRM

To many observers, high technology firms are dynamic, exciting places to work, driven by the quest for continual innovation. Unfortunately, not all jobs in high technology firms are so glamorous (Anderson & Kleingartner, 1987). Wages for many semiskilled production workers and unskilled workers in this industry rank among the lowest in manufacturing industries generally. Many of the jobs themselves are monotonous and boring, with little opportunity for career advancement.

Unskilled, semi-skilled, and clerical workers usually are not offered the array of benefits and incentives that are allocated to the technical and management staffs of high technology firms (Gomez-Mejia & Welbourne, 1990). Such differences in treatment between the "professional" staff and the "nonprofessional" staff potentially can have unfavorable implications for the firm as a whole. They can poison the atmosphere for cooperation and collaboration (Maidique & Hayes, 1984), and they can exacerbate the problems associated with translating innovative ideas into marketable products (Gerpott & Domsch, 1985). Moreover, such differences in treatment create a set of conditions that makes unionization of semiskilled, unskilled and clerical workers in high technology industry more likely to occur. Some observers have pointed to high technology industry as an example of low unionization. Currently that is true, but unless the pay and working conditions of lower-level high technology workers improve, high technology firms may well become targets of union organizing efforts in the not-too-distant future.

In fact, Early and Wilson (1986) identified a list of reasons why some high technology workers might benefit from unionization. These include low wages, inability to deal effectively with grievances, lack of career progression for unskilled employees, low representation of women and minorities, lack of job security and inadequate layoff practices. Proactive managers would be well advised to examine this list critically relative to conditions that exist in their own companies. Doing so now might help improve the productivity and quality of work life of lower-level employees in high technology industry to the point where union representation becomes less attractive.

Despite this possibility, several economic and demographic forces favor a resurgence of unions, and union penetration of the high technology industry. Corporate cost cutting has rattled many workers. A recent survey showed them to be more receptive to unions than at any time since 1979. In addition, labor shortages in certain markets have emboldened workers who no longer fear losing their jobs. And both women and minority-group members, who are expected to continue to enter the work force at a high rate over the next decade, tend to favor unions. However, these same workers are also sympathetic to business. Many of them came of age during the oil shocks of the 1970s, the back-to-back recessions that followed, and then the trade wars. This has led them to appreciate the importance of business in creating jobs and has made them want unions that cooperate with management, rather than confront it (Greenhouse, 1985).

To be more attractive to today's mobile, better-educated, and often white collar workers, unions are changing their tactics. They are bringing in more organizers who are young and female, reflecting the important role that women now play in the work force. They are beginning to understand that while young workers care about wages, many care even more about issues such as day care, job security, quality of work life, developing new skills and having some say about how their jobs are done. Finally, labor is pushing hard to change management's view of labor from a variable cost to a fixed one, through guarantees of job security. This is the wave of the future, and decision makers in high technology firms need to fashion workable solutions to these growing problems, lest they become easy targets for organized labor.

ORGANIZATION DESIGN AND DEVELOPMENT

Here the challenge is to enable line managers to shape and maintain organizational conditions that support innovation and high performance. High performance is synonymous with quality, and quality, as Peters and Austin noted in *A Passion for Excellence* (1986), comes from the belief that anything can be made better. It comes from continuous attention to total customer satisfaction, on the part of every organization member, by providing product quality, services and dependability. Excellence is a game of inches. It's a thousand little things—each done a tiny bit better. Miljus and Smith (1987, p. 127) describe the organization culture of high performance companies:

> A spirit of entrepreneurship is pervasive throughout the internal environments of high performance companies. Every employee, regardless of function . . .. or organizational level . . . is encouraged to actively seek and strive for opportunities to innovate, to contribute, and to be recognized for his or her contributions. The objective is the timely flow of ideas to enhance product value and to improve both product development and product manufacturing processes.

Key design characteristics of these kinds of organizations can be summarized as follows:

1. Relatively flat, project-oriented, matrix structures.
2. Jobs are enriched; employees are empowered to do and to think.
3. There is extensive cross-training and joint problem-solving.
4. Managerial skills and style tend to be participative, with considerable efforts devoted to team building.
5. Organization cultures tend to be egalitarian, collegial, and entrepreneurial. (Odiorne, 1984)

It may be, as Thomas Edison once said, that creativity is 99% perspiration and 1% inspiration. But effective HRM strategies can enhance the likelihood that inspiration will, in fact, occur.

APPLICATION: HEWLETT-PACKARD SINGAPORE

Singapore is one of the jewels among the newly industrialized countries. It is blessed with a tropical climate, a deep water port, an abundance of skilled, semiskilled, and unskilled labor, low wages (by U.S. standards), and a positive balance of payments in world trade. It also is home for over 3,000 employees of Hewlett-Packard. The H-P Singapore plant manufactures integrated circuits (for use in H-P calculators, printers and keyboards); optical couplers (for use in machines that read bar codes, in data communications and in office automation); and light emitting diodes (for use in computer panels, displays and car radios). Computer-aided design and computer-aided manufacturing (including just-in-time inventory control and flexible manufacturing systems) are widely used. The three-story, vertically-designed plant is organized as follows: Floor 1 is dedicated to customer interactions, demonstrations, marketing and sales; Floor 2 is devoted to production, but also includes a large, modern cafeteria and an employee gymnasium (which employees are encouraged to use on their own time); and Floor 3 is devoted to research and engineering, customer hotline assistance (with the capability to simulate most customer problems on computer, for more accurate analysis) and a "failure analysis" laboratory (where products that fail are disassembled to learn why they failed).

However, the most striking characteristics of this operation are not its physical facilities, impressive though they are. Rather, it is the organization's culture—which can best be described as open, team-oriented, egalitarian and participative. Improvement in quality and constant innovation is a never-ending quest. And the signs of it are everywhere.

Workers are organized into semi-autonomous work teams, each of which develops its own performance standards. Each team, along with its supervisor, comprises a quality circle. The names of the teams, including pictures and job titles, are posted on the wall nearest each team. Large, color-coded charts illustrate each team's monthly progress relative to its performance standards.

The teams compete against each other to make suggestions for improvement that actually get implemented. A quality circle in the facilities maintenance group, for example, suggested that to save costs and energy, the company install escalators from floor to floor in the plant, that only run when a person steps on them (otherwise the escalator motor shuts down). The company did so, and was delighted with the results. The company encourages such competition through recognition and rewards. Thus there is competition to become the "top circle." Members of that circle get an all-expenses-paid trip to other H-P plants in other countries to explain their methods, suggestions and innovations. Needless to say, there is heavy peer pressure to do things better.

How are performance standards set? A key input comes from customers, internal as well as external. Customers are surveyed monthly to determine their needs and perceptions. Concurrently, each semi-autonomous work team answers seven questions as part of the process of "Total Quality Control." The following questions are simple to ask, but sometimes difficult to answer.

1. Who are my customers?
2. What are the customer's needs?
3. What results does the customer expect from H-P Singapore?
4. What is our product or service?
5. How do we know if we are exceeding the customer's needs?
6. What is our process for delivering those products or services?
7. What corrective actions are needed to improve?

A picture that helps to facilitate answers to these questions is shown in Figure 1. Together, the inputs of customers, plus the work team's answers to the seven "total quality control" questions, drive performance standards at every level.

All of this does not happen by serendipity. Rather, the HRM process is carefully planned in advance, once a year. Figure 2 shows the overall personnel process for the H-P Singapore plant.

The first step in this process is to try to understand the business for the coming year, as tempered by past trends, local competition and local economic trends. In general, the first step in this process is to understand the country, government, industry and institutional factors that affect the supply of and demand for labor.

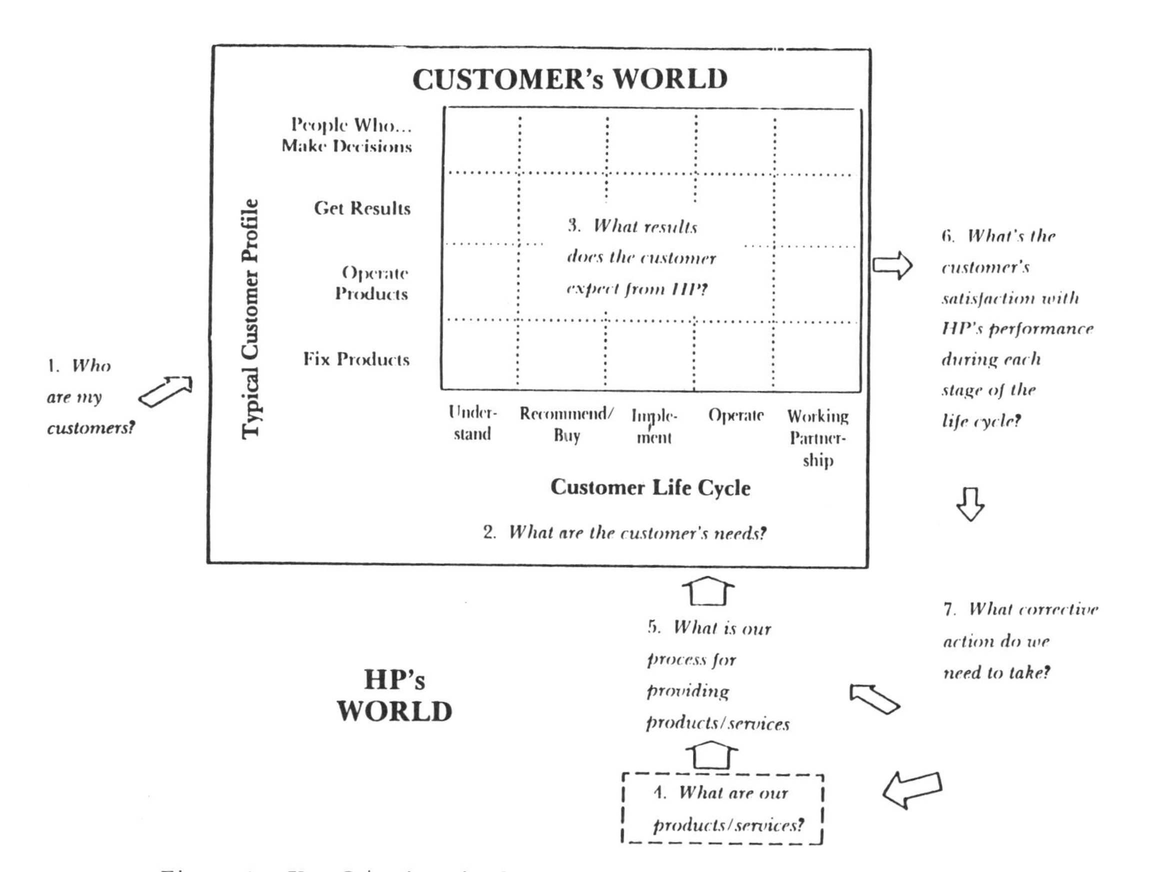

Figure 1. Key Questions in the Process of "Total Quality Control"

Needs analysis then provides input regarding the skill sets required. Such a needs analysis includes: (1) the productivity objectives for each new position; (2) the means for measuring success in the new position; and (3) the compensation and benefits proposed for each new position.

The next step is to carry out staffing resource planning (human resource planning). Needs are fed into training and development, where training strategies are formulated. Training activities are planned one year in advance, with an update each quarter. An estimate of the resources required are then fed back to Staffing Resource Planning, so that H-P can adjust its estimate of the demand for new employees.

In terms of meeting staffing demands, several alternative strategies are possible. New external hires might be brought on board through college recruiting, advertisements, employee referrals, an open-house at the H-P plant or through walk-in interviews. Alternatively, workforce demand could be met by examining existing candidates, coupled with job rotation, overseas assignments or accelerated career development programs. Finally, workforce demands might be met through refitting existing employees by offering skill training, classroom education or interdepartmental loans, and even early retirements.

Once an employee is trained to perform at a competent level, his or her performance then can be appraised meaningfully. Feedback on the new employee's performance is sent to the training and development section so that it might improve its own effectiveness. Pay is related systematically to performance through lump-sum performance bonuses or other incentive programs.

Alternative categories of performers might emerge. "High flyers" and "fast trackers" with "can do" and "will do" attitudes will be rewarded, and characteristics of these people are fed back to the database on skills sets. Poor performers may result from attitude problems or people who cannot fit into available jobs. In an attempt to salvage the situation, refitting and counseling are attempted. Successful attempts are fed back to the database on skills sets. Unsuccessful employees ("no match" in Figure 2) are tracked, and appropriate corrective action is applied. Possibly the compensation for the job is too low, or productivity measures are set incorrectly (too high leads to a condition of "mission impossible," and too low produces mediocre performance).

Once this is accomplished, subprocesses are drafted in support of the key processes. To provide additional information, employee opinion surveys are implemented to determine the perceptions and beliefs of the HR department's key customers: H-P Singapore employees.

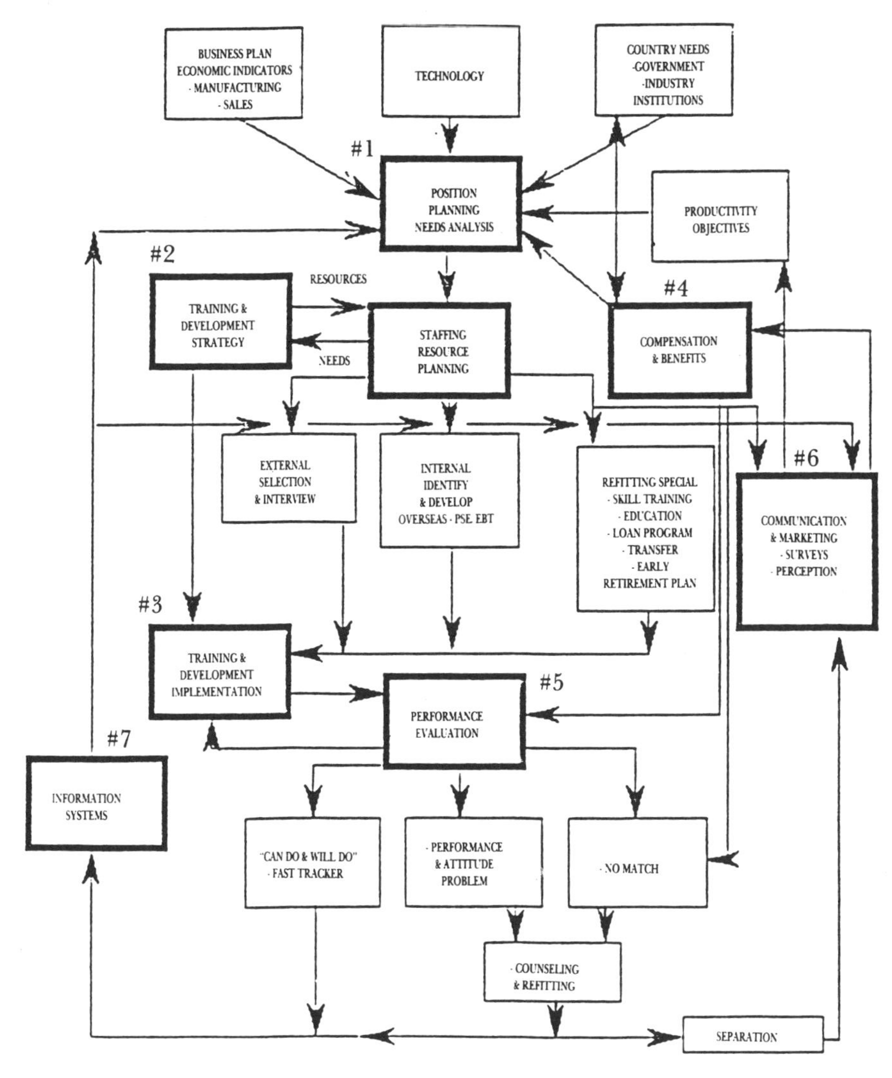

Figure 2. Personnel Process

THE HR DEPARTMENT—DECENTRALIZED THROUGHOUT THE PLANT

One of the most striking characteristics of the H-P Singapore plant is its distributed HR function. HR professionals—called PAMs (Personnel Administration Managers)—are distributed throughout the plant, rather

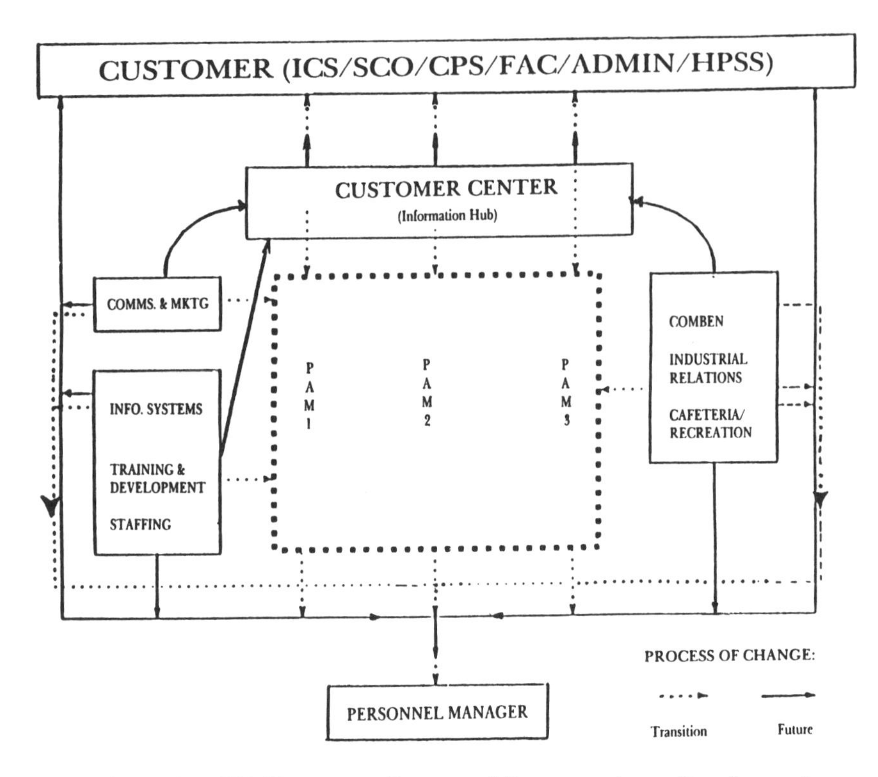

Figure 3. HP Singapore Personnel Structure (next 3 to 5 years)

than concentrated in one department. Each PAM is responsible for all HR issues for one section of the plant. The PAM is accessible, sits in an "open" office, and is easily visible to employees. They get direct, immediate assistance on issues ranging from benefits administration, to vacations, sick leave, payroll and labor relations.

Despite their attractiveness, PAMs are a temporary phenomenon. The vision—HRM of the future—is shown in Figure 3. The function of the PAMs is to counsel, coach and train the line manager who is directly responsible for the operation of a particular section of the plant. The long-range goal is to have operational managers learn to take responsibility for routine HR activities such as:

- accessing employee data bases/information;
- writing performance evaluations;
- carrying out wage reviews;
- forecasting needs for people and skills;
- identifying excess human capacity in their sections;

- formulating with employees plans for their individual training and development; and
- systematically reviewing the results of employee opinion surveys.

The HR operation could then be "downsized," as activities are consolidated into a "Customer Center" or information hub.

ACKNOWLEDGMENT

An earlier version of this chapter was prepared for the first annual conference on Managing the High Technology Firm, University of Colorado, Boulder, January 13-15, 1988.

REFERENCES

Anderson, C.S., & Kleingartner, A. (1987). Human resources management in high technology firms and the impact of professionalism. In A. Kleingartner & C.S. Anderson (Eds.), *Human resources management in high technology firms* (pp. 3-21). Lexington, MA: D. C. Heath.

Balkin, D.B. (1987). Compensation strategies for research and development staff. *Topics in total compensation, 2*(2), 207-215.

Balkin, D.B., & Gomez-Mejia, L.R. (1984). Determinants of R&D compensation strategies in the high tech industry. *Personnel Psychology, 37*, 635-650.

________(1985). Compensation practices in high-technology industries. *Personnel Administrator*, June, 111-122.

________(1987). Toward a contingency theory of compensation strategy. *Strategic Management Journal, 8*, 169-182.

Brody, M. (1987). Helping workers to work smarter. *Fortune, 115*(12), 86-88.

Buffa, E.S. (1984). Making American manufacturing competitive. *California Management Review, 26*(3), 29-46.

Cascio, W.F. (1987). *Applied psychology in personnel management* (3rd ed.). Englewood Cliffs, NJ: Prentice-Hall.

Early, S., & Wilson, R. (1986). Do unions have a future in high technology? *Technology Review, 89*(7), 57-65, 79.

Gerpott, T.J., & Domsch, M. (1985). The concept of professionalism and the management of salaried technical professionals: A cross-national perspective. *Human Resource Management, 24*(2), 207-226.

Gomez-Mejia, L.R., & Balkin, D.B. (1989). Effectiveness of individual and aggregate compensation strategies. *Industrial Relations, 28*, 431-445.

Gomez-Mejia, L.R., Balkin, D.B., & Wilburne, T.M. (1990). Influence of venture capitalists on high technology management. *Journal of High Technology Management Research, 1*(1), 95-107.

Gomez-Mejia, L.R., & Welbourne, T.M. (1990). The role of compensation in the human resource management strategies of high technology firms. In S. Mohrman & M.A. Von Glinow (Eds.), *High technology management* (pp. 440-459). London: Oxford University Press.

Greenhouse, S. (1985, September 1). Reshaping labor to woo the young. *New York Times*, pp. 1F; 6F.

High tech. (1986, June 10). *Wall Street Journal*, p. 1.

Kleingartner, A., & Anderson, C.S. (Eds.). (1987). *Human resource management in high technology firms*. Lexington, MA: Lexington Books.

Maidique, M.A., & Hayes, R.H. (1984). The art of high technology management. *Sloan Management Review*, *26*(2), 17-31.

Manufacturing spending shifts toward training and motivation. (1988, November 10). *Wall Street Journal*, p. A1.

McCarthy, M.J. (1988, April 5). Managers face dilemma with "temps." *Wall Street Journal*, p. 39.

Miljus, R.C., & Smith, R.L. (1987). Key human resource issues for management in high tech firms. In A. Kleingartner, & C.S. Anderson, (Eds.), *Human resource management in high technology firms* (pp. 115-131). Lexington, MA: Lexington Books.

Morano, R. (1973). Determining organizational training needs. *Personnel Psychology*, *26*, 479-487.

Morgenthaler, E. (1986, October 7). Although the scarcities are fewer, some job skills are in short supply. *Wall Street Journal*, p. 37.

National Trades Union Congress News. (1987, October 1). Narrowing the wage gap with other NICs. Singapore: NTUC Press.

Odiorne, G.S. (1984). *Strategic management of human resources*. San Francisco, CA: Jossey-Bass.

Peters, T.J., & Austin, N. (1986). *A passion for excellence*. New York: Harper.

Schroeder, M. (1988, November 7). Watching the bottom line instead of the clock. *Business Week*, pp. 134; 136.

Serving the new corporation. (1986). Alexandria, VA: American Society for Training and Development.

Taylor, M.S., & Bergmann, T.J. (1987). Organizational recruitment activities and applicants' reactions at different stages of the recruitment process. *Personnel Psychology*, *40*, 261-285.

Wanted: Fresh, homegrown talent. (1988, January 11). *Time*, p. 65.

Wexley, K.N. (1984). Personnel training. *Annual Review of Psychology*, *35*, 519-551.

CO-OPTATION AND THE LEGITIMATION OF PROFESSIONAL IDENTITIES: HUMAN RESOURCE POLICIES IN HIGH TECHNOLOGY FIRMS

Ralph C. Hybels and Stephen R. Barley

The popular and managerial press commonly portray high technology as the site of a new corporate order marked by entrepreneurial vision and an enlightened system of labor relations. The general wisdom is that "high tech" firms offer progressive and enviable work environments where management is truly concerned with an employee's welfare. As evidence, commentators note that freedom, autonomy, and financial participation are standard rewards for those who work hard and take risks. In fact, if one believes what one reads, high tech "cultures" embody all the virtues of the long sought for cooperative solution to labor conflict (Walton & McKersie, 1965), a system where workers and management strive in unison for the same organizational goals (Deal & Kennedy, 1982; Peters & Waterman, 1982). There are, however, several reasons to treat such claims circumspectly.

First, the benefits of high technology's progressive and generous approach to employee relations are reserved primarily for a professional elite: the "knowledge workers" with whom high tech's image is so closely associated. Scientists, engineers and other professionals undeniably play a central and critical role in high tech industries, where competition pivots on research and development. However, most high tech firms also employ significant numbers of production and support personnel who often compose the majority of a firm's workforce. These workers ordinarily engage in routine and even temporary jobs that are no less alienating than those found in traditional manufacturing. These production and support personnel rarely enjoy the much lauded privileges awarded to the firm's better educated employees (Burgan, 1985; Gaines, 1988). Membership in the new work cultures may therefore be far more selective than the popular image suggests (Kunda, 1987).

Second, evidence for high technology's cultural uniqueness consists primarily of such human resource practices as flex-time, open door policies, stock ownership, profit sharing, committees for worker participation, educational benefits, job security and an array of firm sponsored recreational activities (see Milkovich, 1987). Although these practices are often touted as innovations introduced by the high technology sector, in point of fact, each originated with firms operating in more traditional industries. Some even harken back to practices that were common at the turn of the century during the era of welfare capitalism (Allen, 1966). Therefore, if employment relations in high technology depart significantly from tradition, it is not because they are particularly unique or original. Rather, what appears to set the high technology sector apart is the extent to which such policies have been adopted concurrently by many organizations.

Third, from a sociological vantage point, the human resource policies typical of high technology firms carry ominous implications for managerial prerogatives and authority structures. For instance, flex-time undermines widely accepted bureaucratic strictures on attendance by means of a myriad of informal agreements that largely allow employees to determine their own hours of work (Perin, 1988). Granting employees the right to participate in project assignments weakens management's prerogative of allocating persons to tasks and, hence, its ability to regulate the firm's internal labor market. Peer review transfers authority to technical employees at the expense of management's control over rewards and punishments. Even practices such as sending employees to conferences or granting leaves of absence expose the firm to a potential loss of proprietary knowledge at the same time that they open the firm to new information. Since powerful groups rarely abdicate control unless they are coerced or unless they seek to gain in one area by trading dependency in another, one must ask whether a whole community of managers would willingly relinquish aspects of authority that are widely held to be their legitimate right simply to build a more egalitarian work place.

Finally, when evaluating high technology's culture one should remember that the history of industrial development has been marked by a succession of managerial philosophies that enjoyed brief periods of widespread acceptance. As Bendix (1956) noted in his classic study of industrial rhetorics, the waxing and waning of managerial ideologies has been associated with shifting social conditions and changing organizational structures. A new managerial ethos is, therefore, likely to signal the importance of a new set of industrial dilemmas and the ideology is, itself, likely to constitute a symbolic response to the practical problems of change (Geertz, 1973; Kunda & Barley, 1988). From this point of view, the sociologically interesting question is not whether high tech cultures represent a significant departure from the past, but rather what organizational dilemmas do they mediate? In other words, why have so many firms found it so convenient, if not necessary, to espouse a more egalitarian philosophy and a progressive stance toward human resources?

ENVIRONMENTAL UNCERTAINTY AND THE DEMANDS OF A PROFESSIONAL LABOR FORCE

Researchers have often observed that the structures of high technology firms are consistent with the "organicism" that contingency theorists claim should be common among firms operating in industries where research and development dominate competitive activity. Burns and Stalker (1961), Lawrence and Lorsch (1967), Thompson, (1967) and others have argued that firms are likely to be less bureaucratic whenever they depend on rapidly changing technologies or exist in turbulent environments. The logic of such an assertion rests on the idea that by maintaining structural flexibility, firms can respond with more agility in the face of uncertainty. From this vantage point, high tech's human resource policies can be construed as part of an environmentally conditioned trend toward "de-bureaucratization."

Contingency theorists have also argued that organic structures are appropriate whenever firms employ large numbers of professionals who must regulate their own work in accordance with occupational as well as organizational guidelines. In organic organizations, a flexible "network structure of control, authority and communication substitutes for hierarchy, consultation largely replaces command, and a technological ethos of material progress and expansion is more highly valued than . . . obedience" (Burns & Stalker, 1961, p. 121). To the degree that professional knowledge cannot be easily rationalized, the organic firm's tendency to attach importance and prestige to forms of control and affiliation valued by the occupation seems particularly adaptive (Scott, 1981).

Most of the popular and academic literature on human resource management in high technology echoes the tenets of contingency theory, not only in its explanation of the origins of high technology's personnel policies, but as a rationale for prescribing their use. High technology firms are said to have adopted progressive human resource policies because such practices foster behaviors and attitudes critical for success in markets that demand high rates of innovation. Proponents of this view assume that autonomy and participation lead to greater inventiveness. For example, Gerstenfeld (1970, pp. 62-65) argues that "porous organizational boundaries," "less conformity," and "an atmosphere promoting individual responsibility and autonomy" are necessary to promote creativity in R&D organizations. Similarly, Glassman (1986, p. 176) counsels that R&D managers should remove what engineers and scientists indicate are obstacles to their creativity, including "lack of time, lack of freedom, and limited communication."

The human resource literature also justifies the practice of granting autonomy and participation as a means of integrating professional values with the norms and objectives of the organization. Professionals have long been portrayed as individuals who are ill-suited for and hostile to the more confining aspects of bureaucratic administration (Kornhauser, 1962; Marcson, 1960). More open organizational structures and limitations on managerial prerogatives are therefore frequently recommended by human resource specialists as a means of accommodating professional expectations and minimizing "professional deviance" (Raelin, 1985; Von Glinow, 1988). The notion is that by altering work environments to make them more "congruent" with professional expectations, managers can prevent disruptive and unproductive clashes between professional and bureaucratic orientations (Von Glinow, 1983). In short, the human resource strategies of high technology firms are openly advocated as a means of co-opting and subsuming professional aspirations under the canopy of an organization's norms and values.

Contingency theories of the origin of high technology's culture are instructive insofar as they lead to a search for explanations based on environmental and technological forces common throughout the sector. In this regard, they seek an ecologically sensitive account for why a whole industry might adopt a progressive stance on its human resources. However, explanations derived from traditional contingency theory cannot explain why the high technology sector has adopted innovative personnel practices while other professionalized industries in turbulent environments, such as banking and health care, have not. The difficulty stems from the fact that contingency theorists envision environments as monolithic entities differentiated solely in terms of abstractions such as degrees of turbulence or uncertainty. The upshot is a form of environmental determinism ill-suited for explaining variability or pinpointing the dynamics by which environments exert pressure (Barley, 1986; Barley & Freeman, 1988).

Moreover, traditional contingency arguments, particularly when used for policy recommendations, bear the trappings of teleological explanations in which effects are treated as causes. In other words, because the practices in question are thought to enhance innovation, commentators assume that firms adopt such practices because they foster innovation. It may well be that by defining responsibilities broadly and by allowing considerable autonomy, organizations can motivate creative involvement (Hackman & Oldham, 1980).[1] It is a mistake, however, to assume that the need to meet market requirements for rapid innovation must therefore be the single most important factor underlying the diffusion of innovative human resource practices in high tech.

To adequately explain why the high technology sector has gravitated so strongly toward a progressive ethos of human resource management, one must first recognize that the environment of high tech firms is multifaceted. These organizations face as much uncertainty in the capital and labor markets as they do in the product market. Some aspects of organizational policy and structure are affected more by conditions in some areas of the environment than in others. To explain why high tech's progressive human resource practices have not been widely adopted by other industries dominated by professionals, it is necessary to identify pressures that are specific to high technology. Finally, to avoid functionalism's teleological trap, one must explore the dynamics of policy making in terms of the particular actions and cognitions employed by the participants.

LABOR MARKETS AND THE SOCIAL CONSTRUCTION OF PROFESSIONALISM

Unlike previous analysts, we contend that the most important determinant of the sector's progressive stance on human resources is the labor market, not the product market. And while we concur that the social dynamics of professionalism are involved, since high technology firms are not particularly unique in their reliance on large numbers of professionals, we take exception with the notion that high technology's human resource practices are an inevitable outcome of the employment of professionals per se. Hospitals, accounting firms, law firms and the chemical industry also employ numerous professionals, yet these organizations are rarely considered to be at the cutting edge of human resource management. The employment of professionals may be necessary, but it is an insufficient reason for the emergence of high technology's cultural practices. To understand why professionalism is insufficient, one must understand why traditional notions of professionalization are inadequate for explaining occupational dynamics among the high tech labor force.

Traditionally, sociologists of work treated the professions as a set of occupations marked by a constellation of special attributes (Carr-Sanders & Wilson, 1933; Greenwood, 1957; Vollmer & Mills, 1966). Professions were said to differ from other occupations insofar as they (1) possess a substantive body of knowledge imparted to novices through systematic training, (2) establish occupational associations that certify and regulate practitioners, (3) enjoy social acceptance of their authority, and (4) adopt a service orientation articulated by a code of ethics. By extension, professionals should subscribe to motives and attitudes that distinguish them from members of other occupations. In particular, professionals should be technical experts who are motivated primarily by autonomy, discretion and work itself; who submit willingly only to the authority of their colleagues; and who work selflessly for the public good. In the 1960s, these ideals gave rise to the notion that professional and bureaucratic modes of organizing should be incompatible (Benson, 1973; Hall, 1967; Scott, 1965). Themes of professional-bureaucratic conflict have, in turn, underwritten much of the literature on scientists and engineers in industry (see Kornhauser, 1962; Marcson, 1960; Pelz & Andrews, 1966; Ritti 1968; Schreisheim, Von Glinow, & Kerr, 1977). As noted earlier, this traditional view of professionalism appears to infuse the rhetoric currently used to justify the human resource practices of high technology firms (Miller, 1986; Raelin, 1985).

During the 1970s, however, occupational sociologists began to question the validity of viewing professions as intrinsically different from other occupations (Freidson, 1970; Johnson, 1972; Larson, 1977; Ritzer, 1977; Roth, 1974). For instance, in his critique of trait-based conceptions of the professions, Klegon (1978, p. 269) suggested that "the characteristics that have often been used to define professions can be best understood as strategies for the achievement and maintenance of a particular type of occupational control." Larson (1977) argued similarly that professionalism is nothing more than a strategy by which occupations eventually monopolize the market for their expertise. Freidson (1970) even claimed that a profession differs from other occupations only insofar as it has been given a state mandated license to control its own work. From this vantage point, it is sociologically more informative to speak of the process of professionalization than it is to debate whether any occupation evidences the attributes of a profession.

Like unionization, professionalization is a strategy for occupational control in which knowledge, education, licensing, lobbying and the politics of reputation are used as ploys for achieving collective power (Child & Fulk, 1982; Van Maanen & Barley, 1984). However, as Hall (1979) points out, power theories of professional action often treat professions as reified actors, thereby obscuring the identity of the parties inside and outside the profession who enact the process. Few theorists have described in detail the specific activities that contribute to professionalization. Filling this gap requires a theory of occupational dynamics. Such a theory should not only explain how occupations that have achieved professional status assert and maintain their power, it should also suggest how occupations that lack full control over their employment situation obtain professional stature.

The claim that a profession employs licensing, training and lobbying in its struggle for self-control assumes that the occupation meets a set of social preconditions that are usually left unarticulated. First, members of the occupation must be conscious of themselves as a distinct group bound together by common interests as well as by differences that are perceived to separate them from other groups. Second, there must exist some avenue of collective action, usually a formal association which presses demands in the name of its members. Finally, society must grant at least incipient recognition of the group's legitimate right to control some domain of expertise.

Surprisingly, these preconditions are fully met by only a handful of occupations, for instance, medicine, law, accounting and funeral directing. In academic science, the precursors to professional power are approximated. But, in engineering and most other technical specialties that populate high tech firms, the preconditions of professional power are poorly developed (Becker & Carper, 1956; Perrucci & Gerstl, 1969; Ritti, 1971). Society, in general, and corporate management, in particular, apparently believe that engineers possess expertise indispensable to the success of commercial ventures. However, it is not clear whether corporations or society are willing to grant engineers full control over their knowledge. In at least some engineering specialties, practitioners continue to be trained on the job under the watchful eye of the organization rather than the occupation (Whalley, 1986). The history of computer science has been marked by an intentional diffusion of programming skills across an ever wider segment of the populace (Greenbaum, 1979). Moreover, no well-established organization can be said to represent actively the interests of engineers in negotiations with other social groups such as governmental regulators or employers (Kornhauser, 1962). The occupational communities of most technical occupations, therefore, remain largely potentialities based primarily on the commonality that members and outsiders (including academics) impute to the motives of workers who pursue seemingly similar activities in relative degrees of isolation.

When occupations lack a formal organization that spans firms, the struggle for professional control must necessarily occur within the confines of the work place since there is no apparatus, such as the American Medical Association or the UAW, for pressing demands at the level of the industry or the society. Under such conditions, the dynamics of professionalization depend heavily on firms' actions. Therefore, rather than view professions and bureaucracies solely as adversaries as has traditionally been the case, it may be more appropriate to consider whether employing organizations might not play a crucial role in the very formation of modern professions. The critical question is what situational dynamics might induce firms wittingly or unwittingly to create and then accommodate the professional aspirations of technical occupations.

In high technology industries, several dynamics appear to have fostered the professionalization of certain technical specialties, the most prominent of which are electrical engineering and computer science. First, a large number of small start-up firms were founded by occupational members who hired other occupational members to serve as executive officers as well as lower level employees (Brittain & Freeman, 1980). In these cases, membership in the organization was largely coterminous with membership in the occupation (Scott, 1965). Managers were themselves likely to espouse occupational norms and aspirations. Second, as contingency theory would predict, when the work of an occupation is central to a firm's day-to-day operation and when there exists no substitute for the occupation's knowledge, firms are more likely to accommodate occupational aspirations. One would, therefore, expect greater occupational influence in organizations such as software houses and biotechnology firms that focus almost entirely on research and development. Finally, and most important, firms are likely to accommodate occupational norms when members of the occupation are in such demand that recruitment and retention become crucial organizational problems.

High technology's rapid growth and the tendency for high tech firms to congregate in specific geographical areas have placed a premium on the availability of computer scientists, electrical engineers, molecular biologists and other specialists crucial to microelectronics and biotechnology firms. The extraordinary rates of turnover and the heavy recruiting requirements of growing firms in the computer industry testify to the sector's difficulty in coping with the labor market (Balkin & Gomez-Mejia, 1984). For over a decade, high tech firms have been forced to compete for a limited pool of skilled specialists. Consequently, the difficulty of obtaining and retaining an adequate supply of talent has gradually elevated human resource management to a more strategically important position than that found in most other industries. From this vantage point, the adoption of progressive human resource policies can be construed as a strategy for coping with a continual scarcity of technical labor.

However, we use the term strategy hesitantly. The degree to which high technology's human resource policies have been consciously and rationally planned is far from clear. As Mintzberg suggests of most strategies, especially those found in the "adhocracies" for which high tech industries are renowned, any specific constellation of practices is likely to emerge gradually from a myriad of related decisions in a variety of firms (Mintzberg, Raisinghani, & Theoret, 1976; Mintzberg & Waters, 1982). Strategy formulation is often less a rational process than a matter of coalition formation, defensive reaction, and imitation (Cyert & March, 1963; Lindblom, 1959; Thompson, 1967). Nevertheless, the human resource practices of the high technology sector do resemble strategies to the extent that they compose a loosely amalgamated family of tactics for maintaining control in the face of a volatile labor market. The intriguing questions are why the tactics bear a family resemblance and why they are sufficiently widespread to give the appearance of a consistent and coherent philosophy of managing professionals. To answer these questions, consider the set of dynamics that might lead managers to the conclusion that they need to formulate such policies.

If managers are to respond to difficult labor market conditions, they must first categorize and make sense of a myriad of employee behaviors, since any employee's decision to change employers always involves a multitude of possible perceptions and motivations (Mobley, 1982). Individual motives may range from the desire for higher pay to a never ending search for excitement. For an organization to address such a variety of motives would require a plethora of contracts and perquisites tailor-made for each employee. Such an idiosyncratic approach would militate against the development of any stance coherent enough to deserve the label of a policy. Consequently, if managers are to understand and address problems of recruitment and retention, they must aggregate data on individuals and formulate responses that apply to those aggregates.

Data on turnover rates and recruitment success in particular departments are often "rolled up" by lower level managers for analysis at an organization's upper echelons. Information on employee motivations gathered from recruitment or exit interviews and employee opinion surveys are also often summarized for policy analysis. Not only are such data frequently categorized by occupation, but when departmental boundaries correspond to functional specialties, even comparisons among departments imply a comparison among occupational groups.

Should the members of an occupational category be widely understood to be seeking professional status, as is the case for engineers, then employers are likely to respond to attraction and retention problems as if they represent demands lodged by a coherent occupational group.[2] What is critical, then, is not whether engineers are professionals or even whether they think of themselves as such, but rather that they are perceived as such by outsiders. By attributing turnover and recruitment problems to cosmopolitan attitudes and other stereotypical attributes of professionals, policy makers can aggregate individual behaviors into collective actions that can be explained by global rather than idiosyncratic motives. Policy responses follow more easily from such analyses. Thus, by reifying occupations as social actors whose influence extends across the boundaries of the organization, managers make recruitment and retention problems more tractable.

The coherence of the high technology sector's human resource strategy derives, therefore, not so much from an enlightened concern with labor relations, as from the fact that managers view their labor problems as rooted in the common needs of large groups of professional employees. Once a model of professional satisfaction becomes a guide for ensuring a steady supply of engineering talent, policymakers are likely to institute practices that cater to the professional's reputed aspirations. If professionals are thought to desire autonomy and participation in decision making, then it would seem logical that they should be attracted to and satisfied by firms that allow greater freedom and discretion (e.g., Von Glinow, 1985). Viewed from this perspective, it should come as no surprise that flex-time, educational benefits, flattened hierarchies, open door policies, stock ownership, participation in project assignments and most other practices typical of high technology are frequently justified by reference to the professionalism of the workforce (Miller, 1986). Such policies may be incrementally implemented and yet appear retrospectively to constitute a coherent industry-wide strategy for dealing with a professional labor force largely because of the dynamics of diffusion and institutionalization.

Because high tech firms compete for the same employees and because successful recruitment and retention in a seller's market is uncertain, new practices arising in one firm are likely to be imitated rapidly by others who wish to appear equally attractive to potential recruits. The upshot is that such practices have rapidly become institutionalized (Dimaggio & Powell, 1983; Meyer & Rowan, 1977; Tolbert & Zucker, 1983; Zucker, 1987). The mimetic diffusion of human resource practices in high technology has probably been accelerated by the fact that computer and biotechnology firms cluster geographically in areas such as the Silicon Valley and Route 128 in Boston. Because interfirm mobility in such areas carries no burden of geographic mobility, industrial clustering exacerbates the risk of a firm being classed as inferior to its competitors simply because it does not offer the same perquisites.

Rather than view high technology's human resource practices as techniques for motivating professional employees, it is probably more accurate to view them as attempts to address problems of recruitment and retention. Since few technical or engineering occupations currently possess the organized power of a fully developed profession, demands for professional status are given voice only through the perceptions and actions of those in positions of influence over organizational affairs. Ironically, however, because human resource policies for engineers are based on the perception that technical employees have professional attributes and motives, they may ultimately legitimate the occupations' professionalism. By awarding engineers and other technical employees participatory rights, educational benefits, autonomy, freedom to determine hours of work and a variety of other privileges in the name of professionalism, human resource managers may actually help spawn the professions they are attempting to control. Specifically, the recipients of the privileges may come to view themselves as having earned their privileges on the basis of membership in particular occupational groups.

In sum, the considerable mobility of high technology labor markets may indirectly foster a process of professionalization in which those who are most involved in furthering the occupation are members of management rather than the occupation's own spokesmen. The extent to which individual engineers and scientists are aware of the consequences of their mobility is unclear and open to empirical investigation. However, even if engineers and scientists do not fully understand how concerted action might contribute to the occupation's control over its work, the perceptions and actions of employers ensure that their individual behaviors will aggregate into a form of collective action.

CONCLUSIONS

We began this paper by asking why firms in high technology industries should have developed such a strong interest in promulgating an enlightened system of labor relations. We noted that the trend was particularly curious since most so-called progressive human resource practices undercut traditional managerial prerogatives and authority. Although it is probable that these practices do create work environments more conducive to innovation, it does not necessarily follow that this has been the primary motivation for their implementation. Consequently, we suggested that researchers need to look beyond received theories in order to understand the development of high technology's approach to human resource management.

At first glance, high technology's human resource practices seem to be a direct outgrowth of the fact that high tech firms employ large numbers of professional and technical workers. However, other industries also have drawn heavily on these populations without actively creating work environments conducive to professional attitudes and perceptions. Moreover, the occupations most prevalently employed by high tech firms possess only the rudiments of the system of professional power enjoyed by physicians, lawyers and other well developed professions. Consequently, if one is to understand why high technology industries have adopted human resource policies that professionalize the work place, one needs to consider the external environment in which the firms exist. In particular, the extraordinary instability of high tech labor markets has driven management to adopt a stance toward labor relations that is conducive to a professional orientation.

Markets for engineers, scientists and other technical workers are not merely professional labor markets, they are markets marked by scarcity and considerable interfirm mobility. Recruitment and retention are therefore crucial problems for high tech firms whose success depends on maintaining an adequate supply of technical talent. These conditions, in turn, encourage policymakers to interpret worker behaviors as motivated by aspirations for professional status and to respond accordingly.

In short, by granting engineers and technical workers autonomy, discretion and other perquisites of professional life, firms hope to create organizational climates that reduce turnover and ease the difficulty of recruitment. Consequently, what at first may seem to be a widespread abdication of managerial prerogatives, appears, on second glance, to be a series of trade-offs aimed at maintaining control. By treating employees as professionals, high tech firms hope to foster the type of loyalty and moral commitment that will bind employees to the firm even when the market offers easy opportunities elsewhere. The strategy of abdicating bureaucratic control in order to gain normative control apparently underlies many of the human resource practices of high tech firms and meshes well with the ideological rhetoric of "organizational culture" that also characterizes the high tech sector (Kunda & Barley, 1988).

We submit, however, that in the long run the new strategies of control may backfire precisely because they create the very conditions they assume. Ironically, high tech firms offer members of technical occupations considerable autonomy, discretion and freedom because they presume that members of such occupations are professionals. Yet, as we have seen, professions are not something that occupations are, but rather something they become.

Engineering and other less well-developed technical occupations have historically lacked the self-consciousness and formal organizational power necessary for pressing claims in the face of employer opposition (Ritti, 1968; Whalley, 1986). However, by granting members of these groups prerogatives in the name of professionalism, employers may unwittingly enhance occupational awareness and legitimate incipient claims to professional status. To the degree that members of technical occupations develop stronger professional self-images, become more conscious of their commonalities, and come to expect treatment as professionals, their loyalty to the occupation is likely to increase. Consequently, by attempting to elicit normative control on behalf of the organization, the new human resource practices may spawn dynamics that eventually enhance loyalty to the occupation. Ironically, then, by attempting to co-opt the motives of professional workers, organizations may find themselves co-opted by incipient professions. As Bendix (1956) so long ago noted, such are the dialectical dynamics of managerial ideologies, the social conditions they seek to address, and the consequences they unintentionally create.

ACKNOWLEDGMENT

This paper has profited from the talents of Gideon Kunda and Pam Tolbert who read and commented on earlier drafts.

NOTES

1. The presumed relationship between enlightened human resource policies and increased innovation is at best complex and has yet to be subjected to careful empirical test (Bailyn, 1985).

2. No doubt, the tendency for the managerial literature to portray engineers as stereotypical professionals contributes to management's proclivity to perceive an engineer's dissatisfactions as expressions of thwarted professionalism. However, as Bailyn (1985) noted, while engineers may desire autonomy and greater influence over organizational decision making, it is by no means clear that these desires reflect a professional orientation. Most employees would prefer more voice in their organizations and educated individuals often perceive voice as an unalienable right.

REFERENCES

Allen, J.B. (1966). *The company town in the American west.* Norman, OK: University of Oklahoma Press.

Balkin, D.B., & Gomez-Mejia, L.R. (1984). Determinants of R&D compensation strategies in the high tech industry. *Personnel Psychology, 37,* 635-650.

Barley, S.R. (1986). Technology as an occasion for structuring: Evidence from observations of CT scanners and the social order of radiology departments. *Administrative Science Quarterly, 31,* 78-108.

Barley, S.R., & Freeman, J. (1988). *Niche and network: The evolution of organizational fields in the biotechnology industry.* Unpublished manuscript, School of Industrial and Labor Relations, Cornell University

Bailyn, L. (1985). Autonomy in the industrial R&D lab. *Human Resource Management, 24,* 29-146.

Becker, H.S., & Carper, J. (1956). The development of identification with an occupation. *American Journal of Sociology, 61,* 289-298.

Bendix, R. (1956). *Work and authority in industry: Ideologies of management in the course of industrialization.* New York: Harper & Row.

Benson, J.K. (1973). The analysis of bureaucratic-professional conflict: Functional versus dialectical approaches. *Sociological Quarterly, 14,* 376-394.

Brittain, J.W., & Freeman, J.H. (1980). Organizational proliferation and density-dependent selection. In J. Kimberly & R. Miles (Eds.), *Organizational life cycles* (pp. 291-335). San Francisco, CA: Jossey-Bass.

Burgan, J.U. (1985). Cyclical behavior of high tech industries. *Monthly Labor Review, 108,* 9-15.

Burns, T., & Stalker, G.M. (1961). *The management of innovation.* London: Tavistock.

Carr-Sanders, A.M., & Wilson, P.A. (1933). *The professions.* Oxford: Clarendon.

Child, J., & Fulk, J. (1982). Maintenance of occupational control: The case of the professions. *Work and Occupations, 9,* 155-192.

Cyert, R.M., & March, J.G. (1963). *A behavioral theory of the firm.* Englewood Cliffs, NJ: Prentice-Hall.

Deal, T.E., & Kennedy, A.A. (1982). *Corporate cultures: The rites and rituals of corporate life.* Reading, MA: Addison Wesley.

DiMaggio P.J., & Powell, W.A. (1983). The iron cage revisited: Institutional isomorphism and collective rationality in organizational fields. *American Sociological Review, 35,* 147-160.

Freidson, E. (1970). *Professional dominance: The social structure of medical care.* New York: Atherton Press.

Gaines, J. (1988). *Changing organizational patterns: A case study of the use of temporary workers at AT&T.* Paper presented at the 83rd Annual Meeting of the American Sociological Association, Atlanta, Georgia.

Geertz, C. (1973). Ideology as a cultural system. In C. Geertz (Ed.), *The interpretation of cultures* (pp. 193-233). New York: Basic Books.

Gerstenfeld, A. (1970). *Effective management of research and development.* Reading, MA: Addison-Wesley.

Glassman, E. (1986). Managing for creativity: Back to basics in R&D. *R&D Management, 16,* 2.

Greenbaum, J. (1979). *In the name of efficiency.* Philadelphia, PA: Temple University Press.

Greenwood, E. (1957). Attributes of a profession. *Social Work, 2,* 45-55.

Hackman, J.R., & Oldham, G.R. (1980). *Work redesign.* Reading, MA: Addison-Wesley.

Hall, R.H. (1967). Some organizational considerations in the professional-organizational relationship. *Administrative Science Quarterly, 12,* 461-78.

________(1979). The social construction of the professions. *Sociology of Work and Occupations, 6,* 124-126.

Johnson, T.J. (1972). *Professions and power.* London: Macmillan.

Klegon, D. (1978). The sociology of professions. *Sociology of Work and Occupations, 5,* 276-292.

Kornhauser, W. (1962). *Scientists in industry: Conflict and accommodation.* Berkeley, CA: University of California Press.

Kunda, G. (1987). *Engineering culture: Culture and control in a high technology organization.* Unpublished doctoral dissertation, Sloan School of Management, MIT.

Kunda, G., & Barley, S.R. (1988). *Designing devotion: Corporate culture and new ideologies of work place control.* Paper presented at the 83rd Annual Meeting of the American Sociological Association, Atlanta, Georgia.

Larson, M.S. (1977). *The rise of professionalism: A sociological analysis.* Berkeley, CA: University of California Press.

Lawrence, P.R., & Lorsch, J.W. (1967). *Organization and environment.* Homewood, IL: Irwin.

Lindblom, C.E. (1959). The "science" of muddling through. *Public Administration Review, 19,* 79-88.

Marcson, S. (1960). *The scientist in American industry.* New York: Harper & Row.

Meyer, J.W., & Rowan, B. (1977). Institutionalized organizations: Formal structure as myth and ceremony. *American Journal of Sociology, 83,* 340-63.

Milkovich, G.T. (1987). Compensation systems in high tech companies. In C.S. Anderson & A. Kleingartner (Eds.), *Human resource management in high technology firms* (pp. 103-114). Lexington, MA: Lexington Books.

Miller, D.B. (1986). *Managing professionals in research and development.* San Francisco, CA: Jossey-Bass.
Mintzberg, H., Raisinghani, D., & Theoret, A. (1976). The structure of unstructured decision processes. *Administrative Science Quarterly, 21,* 465-499.
Mintzberg, H., & Waters, J.A. (1982). Tracking strategy in an entrepreneurial firm. *Academy of Management Journal, 25,* 465-499.
Mobley, W.H. (1982). *Employee turnover: Causes, consequences, and control.* Reading MA: Addison-Wesley.
Pelz, D.C., & Andrews, F.M. (1966). *Scientists in organizations: Productive climates for research and development.* New York: Wiley.
Perin, C. (1988). The moral fabric of the office: *Organizational habits vs high tech options for work schedule flexibilities* (Working paper No. 2011-88). Sloan School of Management, MIT.
Perrucci, R., & Gerstl, J.E. (1969). *Profession with community: Engineers in American society.* New York: Random House.
Peters, T., & Waterman, R.H. (1982). *In search of excellence.* New York: Harper & Row.
Raelin, J. (1985). *Clash of cultures.* Boston: Harvard Business School Press.
Ritti, R. (1968). Work goals of scientists and engineers. *Industrial Relations, 7,* 118-131.
______(1971). *The engineer in the industrial corporation.* New York: Columbia University Press.
Ritzer, G. (1977). *Working, conflict and change.* Englewood Cliffs, NJ: Prentice-Hall.
Roth, J.A. (1974). Professionalism; the sociologist's decoy. *Sociology of Work and Occupations, 1,* 6-23.
Schreisheim, J., Von Glinow, M.A., & Kerr, S. (1977). Professionals in bureaucracies: A structural alternative. In P.C. Nystrom & W.H. Starbuck (Eds.), *Prescriptive models of organizations* (pp. 55-70). New York: North-Holland.
Scott, W.R. (1965). Reactions to supervision in a heteronomous professional organization. *Administrative Science Quarterly, 10,* 65-81.
______(1981). *Organizations, rational, natural and open systems.* Englewood Cliffs, NJ: Prentice-Hall.
Thompson, J.D. (1967). *Organizations in action.* New York: McGraw-Hill.
Tolbert, P.S., & Zucker, L.G. (1983). Institutional sources of change in organizational structure: The diffusion of civil service reform, 1880-1935. *Administrative Science Quarterly, 28,* 22-39.
Van Maanen, J., & Barley, S.R. (1984). Occupational communities: Culture and control in organizations. In B. Staw & L. Cummings (Eds.), *Research in organizational behavior, 6,* 287-365. Greenwich, CT: JAI Press.
Vollmer, H.M., & Mills, D.L. (1966). *Professionalization.* Englewood Cliffs, NJ: Prentice-Hall.
Von Glinow, M.A. (1983). Controlling the performance of professionals through the creation of congruent environments. *Journal of Business Research, 11,* 345-361.
______(1985). Reward strategies for attracting, evaluating, and retaining professionals. *Human Resource Management, 24*(2), 191-206.
______(1988). *The new professionals: Managing today's high-tech employees.* Cambridge, MA: Ballinger.
Whalley, P. (1986). *The social production of technical work: The case of British engineers.* Albany, NY: State University of New York Press.
Walton, R.E., & McKersie, R.B. (1965). *A behavioral theory of labor negotiations.* New York: McGraw-Hill.
Zucker, L.G. (1987). Institutional theories of organization. *Annual Review of Sociology, 13,* 443-464.

INDIVIDUAL, GROUP AND ORGANIZATIONALLY-ORIENTED PERSONNEL SYSTEMS:

IMPLICATIONS FOR STAFFING THE HIGH TECHNOLOGY FIRM

Robert D. Bretz, Jr. and George F. Dreher

The issue of system-congruence is becoming increasingly important in business today. As researchers of the human resource process, we are recognizing the benefits that accrue when an organization's practices are congruent with the underlying strategy of the firm, and the dysfunctional consequences of implementing practices that run counter to the firm's strategic orientation. Currently, however, much of the writing on this topic is speculative and we have yet to provide either an accurate mapping of the strategy-practice domain or empirical evidence to support the prescriptions being made. Given the increasing emphasis on the role of high technology in industry today, an examination of the propriety of specific personnel practices in high technology firms should prove to be not only interesting but also potentially valuable.

To understand the appropriate staffing practices in a high technology firm, we must first arrive at a definition of what a high technology firm is. Kleingartner and Anderson (1987) suggest that the high technology firm is distinguishable by its commitment to and commercialization of new ideas.

They suggest that this concept can be operationalized by (1) a research and development expenditures-to-sales ratio that is at least twice the national average; and (2) a workforce that is heavily weighted in favor of technology-oriented workers. This weight should be at least three times the national average. While high technology companies are generally considered to be in the electronics and/or computing industries, we suggest that they need not be limited to these industries. It seems reasonable to include in the definition of high technology firms, organizations that, through the employment of a higher percentage of engineers and scientists, are leaders in introducing innovation in their industries. This innovation should most likely be in the production process. In this way, the label of high technology is not reserved for a particular subset of industries but includes those companies in all industries that are committed to exploring the frontiers of whatever technologies exist.

THE TRADITIONAL VIEW

Several theorists have provided a framework on which we may base recommendations for staffing practices in high technology firms. Miles and Snow (1978) provide a classification of organizations based in part on the firm's response to its product markets. Some organizations operate in narrow, well defined and stable product markets. These organizations are classified as Defenders. In response to relatively stable and competitive markets, Defenders compete on the basis of efficiency of operations and cost minimization. They seldom take risks by venturing into new or untested markets and as a result, top management tends to be highly expert in its limited market niche. Defender organizations grow by venturing deeper into their current markets.

A very different type of organization operates in rapidly changing, volatile product markets. These organizations are actively seeking new market opportunities and as such are referred to as Prospectors. Prospectors compete on the bases of innovation, technical superiority and being first to market with new products. Because of the emphasis on speed and change, these organizations are not as efficient as they could be. These organizations are characterized by a willingness to take risks in their product markets and a propensity to search for new markets to exploit. These are the organizations that are most likely to be considered high technology firms.

A third type of organization, the Analyzer, operates in both relatively stable and unstable product markets. These organizations can be described in terms of the Prospector or Defender by changing the unit of analysis from organization to business unit. Those units that operate in stable environments behave much like Defenders, while those operating in unstable environments take on the characteristics of Prospectors.

Olian and Rynes (1984) present a series of speculative propositions about the propriety of various human resource practices in Defender-like or Prospector-like organizations. They limit their discussion to differences in staffing systems that are likely to be observed as a function of strategic type, and base their propositions on two key assumptions: "(1) different types of organizations require different types of people (especially at the managerial levels) for effective performance, and (2) different recruitment and selection practices attract different types of individuals into organizations" (pp. 170-171).

In particular, Prospectors are expected to need and be attractive to potential employees with backgrounds in research and marketing. Because of the organization's propensity to pursue new and untested waters, individuals with high tolerance for ambiguity and willingness to take risks are preferred. As a result of the rapidly changing nature of the organization, selection criteria will most likely focus on past achievements. These organizations should recruit and hire at multiple levels and may rely on less formal selection procedures.

Drawing on what appears to be Porter's (1980) classification of organizations, Schuler and Jackson (1987) offer suggestions as to the appropriateness of personnel practices in organizations employing different strategies to obtain a competitive advantage. Schuler and Jackson suggest that when deciding on which personnel practices to implement, organizations should consider which employee behaviors best contribute to the implementation of its chosen strategy. Following Porter's model, an organization can achieve a competitive advantage in one of three ways. Under a Cost-reduction strategy, an organization gains a competitive advantage by being the lowest cost producer of a good or service. Under a Quality-enhancement strategy, an organization gains a competitive advantage by focusing on product improvement and zero-defect operation. Under an Innovation strategy, an organization gains a competitive advantage by developing and marketing new products and services. The strategy of Innovation is most consistent with the current discussion of high technology firms.

An organization following an Innovation strategy would be concerned with encouraging a high degree of creative behavior from its employees, a longer-term focus and considerable cooperative, interdependent behavior. It would also want to hire those with a high tolerance of ambiguity and greater propensity to take risks (Schuler & Jackson, 1987, p. 209). From the traditional viewpoint, staffing practices that encouraged these types of behaviors would appear to be reasonable practices for a high technology firm.

Strategists have also attempted to classify organizations in regard to where they are in their product's life cycle. Such a framework is provided by

Gerstein and Reisman (1983). This framework includes strategies for organizations in the following phases of product life cycle:

1. Entrepreneurial;
2. Dynamic Growth;
3. Extract Profit;
4. Liquidation/Divestiture; and
5. Turnaround.

The phase of the life cycle that is most consistent with a high technology orientation is the Entrepreneurial stage. The practices implemented to facilitate this strategy should encourage innovation, risk taking, cooperation with others, long-term focus and commitment to the organization (Schuler, 1986, 1987). Therefore, staffing practices that accomplish these ideals appear to be consistent with what a high technology firm would also want to accomplish.

Kochan and Barocci (1985) have also suggested how human resource decisions can be congruent with stages in the organization's life cycle. The Introductory stage typically occurs when the organization is young and is expanding to the point that the founder has trouble managing it alone. The Growth stage is the period during which the organization's products are becoming better known in the market and exposure is increasing. The Maturity stage brings the organization an established position in its market niche. The Decline stage is characterized by changing markets and organizational adjustments to survive in the new environments. Organizations in an Introduction or Growth stage may operate in a fashion similar to Prospectors and as such display a high technology profile, while those in a Mature or Declining stage may function more like Defenders.

Organizational characteristics and the traditional view of staffing and succession patterns for managerial, professional and technical employees are outlined in Table 1 for the organizational typology and life cycle stage approaches.

A SYSTEM-CONGRUENCE VIEW

The concept of strategy-practice congruence is a common thread running throughout this literature. Theorists seem to agree that an organization can choose many routes to success, as evidenced by the menu of strategic options available to it. However, virtually all theorists agree that only by successfully implementing the practices that are congruent with the chosen strategy will the organization flourish. When incongruent strategy-practice linkages are attempted, the result is hypothesized to hinder the organization in the pursuit of its stated goals.

Table 1. Organizational Characteristics and Staffing/Succession Pattern for Managerial, Professional, and Technical Employees: The Traditional View

Organization Type[a]	
Prospector	*Defender*
To be able to move quickly into new businesses, managerial, professional, and technical talent is often acquired from outside the organization.	Individuals tend to enter at low levels, receive considerable on-the-job training and steady promotions if they are of high potential. This is possibly due to more highly centralized organizational structure and functional orientation. Managers tend to have narrow specialized skills. This is appropriate since the overall emphasis is on the development of efficient technology and production.
More likely to be more decentralized and organized along divisional or product lines. Employees likely to face relatively frequent changes in duties and assignments due to changing strategic direction.	
Upper-level hiring decisions focus on selection criteria related to proven achievement and specified levels of knowledge, skills and ability.	Selection criteria give minor weight to past achievements and instead focus on future aptitude and potential.
Less likely to formalize selection criteria since job requirements are revised with changes in strategic direction.	Due to stability, types and levels of job qualifications are clearly articulated. Selection and promotion criteria are formalized.
Individuals with a preference for risk taking and unstructured environments best suited for this organization type.	Individuals with high needs for security and structure and low tolerance for change and ambiguity best suited for this organization type.
General reliance on external sources of managerial talent.	Managers tend to be promoted from within. General reliance on internal versus external recruits.
Increased need to use external search and recruiting agencies.	Reduced need to use outside search and recruiting agencies.
More likely to use selection devices that emphasize work history.	More likely to use selection devices that assess applicants' future aptitudes and potential promotability.

(continued)

Table 1. (Cont'd)

Life Cycle Stage[b]	
Introduction—Growth	*Maturity—Decline*
Recruitment and selection of technical and managerial personnel who already possess skills needed to make product successful are stressed. The objective is to recruit the best talent available. Training and employee development are likely to receive less attention, since hiring people who already have required skills is more cost effective. As firms move into the growth stage, they define future skill requirements and begin to establish career ladders. By trying to stay ahead of competition in product innovation, priority is given to recruiting up-to-date technical talent.	The key staffing issue becomes how to cope with a labor force that is larger than necessary. Priority is given to minimizing compensation costs and staffing levels since maximizing productivity and improving efficiency are of major concern. Well developed career ladders are likely to enhance ability to match scarce job openings with talent available within the organization. As firms move into the decline stage, they plan and implement workforce reduction and reallocations. The implementation of retraining and career consulting services characterize firms in the decline stage.

Notes: [a] This view of strategy-staffing practice linkages is derived from Olian and Rynes (1984).
[b] This view of organizational context-staffing practices linkages is derived from Kochan and Barocci (1985).

Staw (1986) offers a framework for combining human resource practices into logical "systems" that when implemented in unison can powerfully affect the direction of the organization. His ideas are based on the notion that not only must strategy and practice be congruent, but practices must be congruent with one another. The purpose of all of the systems he describes is to improve the overall performance of the organization through employee satisfaction and performance. However, the three systems attempt to accomplish this objective in very different ways. The Individually-oriented system recognizes the importance of rewarding individual behavior. The rationale underlying this approach is to structure the reward system so that the instrumentality linkage between rewards and individual performance is explicit. Rewards should be tied to performance. Performance should be appraised on the basis of individual skill and/or output. Jobs should be designed to allow individual contribution. Training should focus on building the skills needed to achieve valued rewards. Realistic and challenging goals should reinforce the expectancy linkage.

Table 2. Staw's Three Systems of Organizational Change

Individually-oriented System

Extrinsic rewards linked to the performance and contribution of individual employees.
Realistic/challenging goal setting.
Individual employee performance appraisal and feedback programs.
Skill and performance-based promotion system rather than sponsored mobility.
Job design to increase responsibility, variety, and significance.

Group-oriented System

Work organized around intact groups.
Group members participate in the selection, training, and rewarding of members.
Groups enforce the norms of behavior.
Resources and rewards are distributed on a group rather than individual basis.
Intergroup competition is encouraged.

Organizationally-oriented System

High degree of organization-wide socialization.
Job rotation (to encourage company-wide loyalty).
Long-term, company-specific training.
Long-term, protected employment.
Decentralized operations.
Sharing of financial and strategic information about the firm.
Rewards linked to organizational performance, such as, profit sharing, stock options.

Source: Staw, 1986, pp. 40-53.

The second system that Staw describes is Group-oriented. The rationale underlying this approach is to structure the reward system such that group participation and performance are the primary determinants of reward allocation. Since rewards are determined at the group level, the group is able to exercise control over the members. This approach requires designing jobs around intact groups such that interaction is required to successfully complete the work. Since work is done cooperatively, performance is appraised on the basis of group contribution. Rewards are distributed to the group which then decides how the within-group allocation will take place. Groups should be relatively autonomous, with control over selecting and training members. Controlled competition between groups may be desirable.

The third system is Organizationally-oriented. Here, the underlying rationale is to structure the reward system such that individuals gain satisfaction and rewards from contributing to the welfare of the entire organization. This requires a strong socialization effort to develop organizational commitment. Job rotation would allow individuals to develop an organizational perspective and keep them from becoming tied to a particular unit or subunit. Training programs should focus on providing firm-specific training that raises the individual's value within the organization but is not easily transferrable to other organizations. Long-term employment contracts or implied contracts should increase organizational loyalty. Decentralized operations and few status distinctions would help remove dissent and feelings of separation. Individual rewards, at all organizational levels, should be tied to organizational performance. Profit sharing, bonuses and stock options should be utilized. The characteristics of Staw's systems are summarized in Table 2.

STAFFING PRACTICES FOR A SYSTEMS APPROACH

The staffing practices suggested by each of Staw's systems are different and have different implications for the organization. The general objective of the Individually-oriented system is to create a high degree of individual motivation and entrepreneurial behavior. Therefore, staffing practices appropriate in this system should emphasize individual ability and achievement. This will most likely be accomplished by relying on a contest mobility system in which employees compete for promotions and raises. Screening and selection processes should focus on past achievement and proven ability. Individuals who are achievement oriented, willing to incur risks, and have a strong desire for autonomy would seem to be best suited to this type of organizational climate.

In an Organizationally-oriented system, the general objective is to create a high degree of commitment and loyalty to the firm. In that regard, a sponsored mobility system that channels employees into particular career ladders on the basis of some predetermined criteria may be observed. Screening and selection processes would likely focus on aptitude and measures of cognitive ability rather than on past achievements. Staffing decisions may hinge on the degree to which the applicant is viewed as offering potential for long-term contribution to the organization in many possible contexts. Individuals who desire interaction, are risk averse and have higher affiliation needs seem better suited to this type of organization. Staffing decisions associated with Individually-oriented and Organizationally-oriented systems are summarized in Table 3. The Group-oriented system is not considered here since it shares characteristics with the other two systems and may confuse the comparison.

THE APPARENT CONFLICT

The prior discussion has suggested that high technology firms are generally classifiable as Prospectors, Entrepreneurial, Innovative or Growth-oriented.

Table 3. Staffing Decisions Associated with Individually Versus Organizationally-Oriented Personnel Systems

Staffing Decisions	*Individually Oriented*	*Organizationally Oriented*
Mobility System	Contest	Sponsored
Recruiting emphasis	External/internal	Internal
Screening/selection procedures (capacity)	Samples (Work samples, assessment centers, situational interviews)	Signs (Measures of aptitude, cognitive ability)
Screening/selection procedures (willingness)	Autonomy, achievement, willingness to take risks (Standard personality measures)	Affiliation, right types (Empirically-keyed personality measures, bio-data)
Retention	Retain top performers	Create a general climate of job security

With the possible exclusion of the Prospector typology, these types of organizations generally emphasize the importance of a high level of cooperative, interdependent behavior, a longer term focus and commitment to the organization. The practices implemented to achieve these objectives are consistent with those suggested by an Organizationally-oriented approach. Herein lies the problem. There is reason to suspect that the kinds of employees needed by high technology firms will find Individually-oriented systems more attractive and Organizationally-oriented systems incongruent with their internal need structures.

One would expect high technology firms to employ a larger percentage of scientists and engineers than other organizations. In fact, this is one of the identifying features of high technology organizations. Since a staffing decision requires two choices rather than one (the organization chooses the person and the person chooses the organization), it is reasonable to consider what type of organization engineers and scientists are likely to prefer.

On this topic there is only limited empirical evidence. Siess and Jackson (1967) report correlations between measures on the Strong Vocational Interest Blank (SVIB) and those on the Jackson Personality Research Form (PRF). The current discussion suggests that an Individually-oriented system will be preferred by those individuals exhibiting needs for achievement and autonomy while an Organizationally-oriented system should appeal to those exhibiting higher need for affiliation. Therefore, given the systems in place in most high technology firms (Organizationally-oriented), if these firms are to appeal to scientists and engineers, then these individuals should be low on need for achievement and autonomy and high on need for affiliation. In reality, the evidence provided by Siess and Jackson (cited in Jackson, 1984) suggests that the opposite is more likely true.

Of the vocational classifications generated by the SVIB, those most applicable to high technology firms are biological scientists, chemists and physicists and engineers. In all cases, these vocational interest categories show moderate positive correlations with PRF measures of need for achievement and need for autonomy, and moderate negative correlations with need for affiliation (see Table 4). This suggests that the very people that high technology firms need to attract find the organization's environment to be incongruent with their need structures. These individuals seem better suited to organizations that reward individual performance through both contest mobility systems and clear instrumentality linkages (i.e., Individually-oriented systems).

Table 4. Correlations Between Strong Vocational Interest Blank Classifications and Jackson PRF Need Strength Measures

	Need For		
Vocational Classification	*Achievement*	*Affiliation*	*Autonomy*
Biological Scientists	.29	—.06	.10
Chemists and Physicists	.25	—.17	.17
Engineers	.27	—.14	.11

Source: Siess and Jackson (as cited in Jackson, 1984, p.49).

This presents a dilemma for high technology firms. On the demand side, since these are often small or fledgling companies, they often rely on promises of deferred compensation (e.g., profit sharing, stock options) in order to build their capital base and remain viable in the early years. However, their success is in many ways driven by the supply of scientists and engineers willing to provide their services. Given the demand for these individuals, and the bidirectional nature of the staffing decision, it does not appear that firms can long ignore the need structures of these key individuals. To the extent that job applicants have a choice when deciding where to work, basic motivation theory (e.g., Murray, 1938) and the vocational choice literature (e.g., Holland 1973; Super, 1953) suggest that they will choose an environment congruent with their need structure and one in which the expression of these needs is allowed and rewarded. Given this, high technology firms may consider deviating from the traditional view and providing their research scientists and engineers with reward and mobility systems consistent with high need for achievement and autonomy. Perhaps characteristics of an Individually-oriented system, such as contest mobility norms and strong individually-based instrumentality linkages, can be introduced into high technology firms that will enhance perceptions of person-environment congruence for these key individuals and at the same time maintain the strategy-practice congruence necessary for long-term survival and organizational health.

REFERENCES

Gerstein, M., & Reisman, H. (1983). Strategic selection: Matching executives to business conditions. *Sloan Management Review, 24*(2), 33-49.

Holland, J. (1973). *The psychology of vocational choice* (rev. ed.). Waltham, MA: Blaisdell.

Jackson, D. (1984). *Personality research form manual.* Port Huron, MI: Research Psychologists Press.

Kleingartner, A., & Anderson, C. (1987). *Human resource management in high technology firms.* Lexington, MA: Lexington Books.

Kochan, T., & Barocci, T. (1985). *Human resource management and industrial relations.* Boston, MA: Little, Brown.

Miles, R., & Snow, C. (1978). *Organizational strategy, structure, and process.* New York: McGraw-Hill.

Murray, H. (1938). *Explorations in personality.* New York: Oxford University Press.

Olian, J., & Rynes, S. (1984). *Organizational staffing: Integrating practice with strategy.* Industrial Relations, *23*(2), 170-183.

Porter, M. (1980). *Competitive strategy.* New York: Free Press.

Schuler, R. (1986). Fostering and facilitating entrepreneurship in organizations: Implications for organization structure and human resource management practices. *Human Resource Management, 25*(4), 607-629.

________(1987). Human resource management practices and organizational strategy. In R.S. Schuler, S. Youngblood, and V. Huber (Eds.), *Readings in personnel and human resource management* (3rd ed., pp. 24-39). St. Paul, MN: West.

Schuler, R., & Jackson, S. (1987). Linking competitive strategies with human resource management practices. *Academy of Management Executive, 1*(3), 207-219.

Staw, B. (1986). Organizational psychology and the pursuit of the happy/productive worker. *California Management Review, 28*(4), 40-53.

Super, D. (1953). A theory of vocational development. *American Psychologist, 8,* 185-190.

STAFFING ISSUES IN THE HIGH TECHNOLOGY INDUSTRY

Mark S. Turbin and Joseph G. Rosse

Attracting and retaining scientists and engineers is a major concern in many organizations. In the popular press, trade publications, and the academic literature, one finds reports that high technology companies spend huge sums for recruiting but often fail to meet their needs for scientists and engineers, who are essential to an industry where competitiveness depends on research and development (R&D) of new products. The supply of these high technology professionals is limited and the demand is increasing because many high technology organizations are young and growing, while firms in more traditional industries also are adopting high technology processes. This supply-demand imbalance makes recruiting new members of the labor force competitive and allows a great deal of mobility for scientists and engineers who are already employed.

The high technology management literature is replete with statements that reflect two assumptions: (a) certain characteristics of the work and the labor market in the high technology industry are responsible for creating and/or exacerbating difficulties in attracting and retaining scientists and engineers; and (b) many of these characteristics are unique to high technology industry. The purpose of this paper is to review what is known regarding those assumptions, and strategies for addressing those difficulties, and to clarify what we still need to learn in order to improve in this area of human resource management. This paper is based on information from the nascent high technology management literature, much of it anecdotal, and from interviews with human resource managers in seven high technology organizations in the vicinity of Boulder, Colorado. We report the available data with confidence that they are accurate descriptions of the observers' experience, but might not generalize to other situations. In this same spirit, the ideas and conclusions presented herein should be considered tentative, and useful for generating further ideas rather than providing conclusive evidence.

THE HIGH TECHNOLOGY ENVIRONMENT

Human resource management practices must accommodate certain typical characteristics of high technology organizations and the labor market in which they operate. A definition of high technology firms given by Milkovich (1987) describes a general consensus of relevant features of this industry: "firms that emphasize invention and innovation in their business strategy, deploy a significant percentage of their financial resources to R&D, employ a relatively high percentage of scientists and engineers in their work force, and compete in worldwide, short-life-cycle product markets" (p.271).

Because of the large number of small high-technology companies and the short life cycles of products, there is intense competition to bring products into production and distribution very quickly. A lack of engineers can cause product ideas not to be implemented, and any delay is a serious threat to market share. Therefore R&D personnel are a crucial resource for the survival of high technology firms, which compete for this resource not only with one another, but also with academic research centers and other manufacturers converting to computer-controlled and robotics-based operations (Miljus & Smith, 1987).

In one sense, the demand for scientists and engineers is increasing faster than the supply. The supply of new college graduates might be as large as the number of people needed, but they typically do not have the specialized skills and experience high technology companies want. Severe shortages in particular specialties such as radar or software engineers may result in a position remaining vacant for months at a time before the right person is found. Hiring for such positions is a major investment; recruiting costs alone have been estimated at $5,000 to $7,000, and nearly twice that if a recruiting agency is used (Seelig, 1985). Companies are reluctant to risk hiring someone not well suited for the position. Another shortfall in some locations is the hiring of women and minorities, creating problems for affirmative action. Thus, despite urgent pressure to fill vacant positions, the number of recruits available is not as great a concern as attracting the right person (McCarthy, Spital & Lauenstein, 1987).

Besides particular specialties, experienced engineers also are in short supply. Those with two to five years of experience are in demand because they have proven ability, but still have reasonable salary expectations (*"Business woos engineers,"* 1979). There is also a great demand for engineers with six to ten years of experience to direct projects (McCarthy et al., 1987; Parden, 1981). It is hard to meet this need, partly because many engineers and engineering students switched to other fields when hiring in defense and space industries was down around 1970.

The most obvious source for experienced scientists and engineers is other high technology firms, and "raiding" is common, especially among companies that are concentrated in an area where property values act as an obstacle to relocation, such as the Santa Clara Valley of California and the Route 128 corridor of Boston. Seelig (1985) cited industry data showing that companies lose one third of newly graduated engineers within three years of hiring, with overall turnover rates from 17 to 25%. Our interviewees reported that it is common for scientists and engineers to change companies after four or five years, and most reported turnover rates between five and thirteen percent. (An exception to this mobility is that some scientists are so specialized that there is little demand elsewhere for their skills, even though the supply is small.) There is severe salary compression for scientists and engineers as new hires get external market rates, so moving to a different organization is believed to increase one's salary. There is little risk of prolonged unemployment if the new company fails or the job does not meet expectations, and there seems to be less stigma than in the past attached to changing organizations. It is not uncommon to return to a company one previously left.

Another relevant feature of the high technology environment is the nature of scientists and engineers themselves, who often are thought to be different from other employees in many ways. Golson (1985) described a technically oriented personality. Technical professionals, he said, tend to be highly intelligent, serious, socially reserved and overly wrapped up with their own ideas and projects. Other typical elements of this image include very positive self-concepts, poor interpersonal skills, high achievement orientation, strong concern for career advancement and discomfort in making quick decisions on the basis of sketchy information.

Career advancement in high technology industry often is equated with management rather than a technical career. Technical specialists have limited opportunity for advancement via increased responsibility and supervision of others. Many engineers in their mid-thirties go back to school for MBAs, seeing management as the only way to advance their careers (Kail, 1987). Management positions provide decision-making latitude, credit for achievements and control of budgets for staff and equipment (Greenwald, 1978). These factors may be important in retaining talented personnel.

Greenwald (1978) reported that only 19% of professionals who had spent time in management were dissatisfied with their career choices, compared with 46.9% of those who had not managed. Small, new companies are attractive because executive management is only a few steps from entry level professional positions. Those with entrepreneurial interests often leave a company to start one of their own.

Some scientists and engineers do not desire management duties and responsibilities, so this feature of the high technology industry presents a potential for dissatisfaction with work and career advancement. However, if a high technology company is growing rapidly, it may be able to provide more career opportunities in both management and technical areas.

OBSTACLES TO ATTRACTION AND RETENTION

The characteristics of the high technology business environment described above suggest several possible causes for difficulties in attracting and retaining scientists and engineers. Such causes are summarized below, grouped according to the issue to which they were most often linked in the literature and interviews, although these issues undoubtedly share some common causes. In addition to categories for attraction and retention, the issue of productivity is also included. Even though productivity seldom was mentioned in connection with difficulties in maintaining adequate staffing levels, improving attraction, retention and productivity, all provide means of increasing the amount of work that can be accomplished. Productivity enhancement may be especially important for firms that are no longer growing (Grissom & Lombardo, 1985).

Attraction

The most conspicuous feature of the high technology industry that provides a challenge to attracting scientists and engineers is the tight labor market. Job seekers face offers of interesting work and lucrative compensation from virtually all the organizations competing to recruit them. Thus, employers must go beyond those standard elements to make one offer stand out from the others. This can be particularly troublesome to small startup companies with limited capital.

One aspect that makes recruiting harder is relocating employees and their families. In some areas where high technology firms are concentrated, the cost of living has been driven up, requiring correspondingly high salary offers and relocation benefits. On the other hand, a lack of competing employers also creates difficulties in recruiting employees who are hesitant to move to a location that offers few alternatives in case the job does not meet expectations. Other barriers to relocation include a spouse's career and moving children to new schools.

Ironically, a favorable geographic location is often mentioned as a recruiting advantage, despite the problems associated with a concentration of high technology firms. Many high technology centers are characterized by pleasant climate, aesthetic surroundings and cultural and recreational opportunities. Access to continuing education, the quality of local university programs, state of the art technology in the community and an atmosphere of support for entrepreneurial activity offer good opportunities for professional growth and development. Some areas suffer a comparative disadvantage from a lack of cultural diversity, especially in regard to recruiting minorities.

Another challenge to the job of recruiting is the uncertainty that makes it difficult for many companies to promise job security. The high technology industry is so competitive, most new startup companies fail to survive. In addition, those that survive suffer periods of retrenchment as the industry goes through cycles of good and bad times, especially companies that depend on government contracts for a substantial part of their business. A record of strong growth and quick recovery from downturns provides a recruiting advantage. Uncertainty also can be offset to some extent by the presence of other suitable employers in the immediate vicinity, so that a job change will not necessitate moving to a new home.

Retention

Demand for scientists and engineers with at least two years of experience provides ample opportunity for mobility. These labor market conditions amplify the level of turnover that might normally result from influences such as financial rewards, career advancement and the nature of work assignments. "Salaries are the major, tangible, factor in job hopping. . . . Opportunity for advancement, to grow and develop as professionals, and to do challenging work, are the key intangibles in the technical specialists' decision to move within or outside of their organization" (Parden, 1981, p. 2).

Other causes of turnover have to do with management in high technology firms. Management duties typically are assumed by scientists and engineers who have little or no management training, and who often are perceived as lacking in interpersonal and communication skills. Too often, providing feedback and recognition is neglected. Dissatisfaction with working under such management is widely reported to contribute to turnover (Parden, 1981; Sherman, 1986). The need for management development was a common theme in our interviews, yet in this fast paced industry many employers—especially smaller companies—find it difficult to muster time and resources for such activities.

Scientists and engineers sometimes become dissatisfied about being required to assume management duties. Many prefer to avoid the burden of administration and supervision, or feel uncomfortable making daily management decisions without the luxury of complete information and analysis. Unless a company is quite large or rapidly growing, there are few high level nonmanagement positions, so opportunities for advancement are very limited for those who choose not to supervise others.

Productivity

Productivity of scientists and engineers is said to suffer from a lack of training in critical nontechnical skills, such as writing, oral presentation and coordinating one's work with others as a team member. Managers' failure to give adequate feedback and recognition contributes to misunderstanding of priorities and misdirection of work efforts.

Productivity also can suffer from time spent by technical professionals on tasks that do not demand their specialized knowledge (and high salaries). Through oversight or the belief that the company cannot afford to hire support staff, scientists and engineers often do work that could be performed by technicians, clerks or secretaries.

Errors in selection and placement when scientists and engineers are hired also can reduce productivity. Technical specialists who start small firms often are suspicious of professional managers and may resist delegating hiring and placement responsibilities to human resource specialists. Filling a position with a person who does not have the proper training or experience, while tempting in a tight labor market, has detrimental effects on productivity, terminations and voluntary turnover.

CURRENT PRACTICES

High technology organizations exhibit a variety of innovative approaches in response to these obstacles to attraction and retention. Kochan and Chalykoff (1987) noted that innovation in human resource management is correlated with tighter labor markets and more high-skill, managerial, technical and professional occupations. However, this may be confounded with growth stage of the organization. Milkovich (1987) reported that new, growing companies (high technology or not) typically exhibit more innovative practices in the design of pay systems, and the same may be true of other human resource management practices as well.

Attraction

Young companies with limited capital typically offer base salaries below market, but offer large bonuses and shares of stock (Balkin & Gomez-Mejia,

1985). Ownership in a startup company before it goes public can lead to tremendous wealth if the company prospers. High technology firms tend to emphasize incentives in the compensation packages of a greater proportion of their technical and managerial work force (Milkovich, 1987). This allows smaller companies to offer job candidates total compensation packages that compare favorably with the higher salaries offered by larger, more mature companies. Stock options and pension plans vested over several years can be offered in hopes of improving both attraction and retention.

Other inducements that are used to improve recruiting include prospects for promotion, leading edge work and travel to overseas locations. Companies with smaller technical staffs might be able to offer more task identity and variety. Some companies use job security as a recruiting aid by retraining or transferring employees to maintain a reputation for avoiding layoffs. Hiring also is facilitated by recruiting from the company's local area to minimize relocation expenses.

To facilitate recruiting, some larger companies commit considerable resources to improving college relations. As one example, Carnegie-Mellon University increased its funding from private industry by a factor of 24 in 5 years ("Wanted: High Tech Engineers," 1985). By offering internships, scholarships, equipment and grants to schools, companies enhance their images in the eyes of prospective employees while improving the quality of new graduates. Some companies target particular universities with known good programs for a majority of their new graduate hires.

Retention

In order to improve retention, companies have responded to technical professionals' desires for development and career advancement. Keeping employees in touch with state of the art developments is beneficial for performance as well as satisfaction and retention. Some firms have formal development plans to specify long and short term goals and the training and experiences needed to reach them. Even though typically high turnover rates lower the return on investment in training, companies have offered in-house training, continuing education, seminars and as much as 5 hours a week release time to pursue advanced degrees. One firm described a development strategy that reduced employee mobility. This involved hiring individuals who had not completed a terminal degree, and providing internal training not leading to a degree, which allowed employees to advance beyond positions for which other employers would hire them.

Dual career ladders, in which technical positions are paid comparably to managerial assignments, can reduce turnover among engineers and scientists, who would otherwise have to choose between management careers they do not want or a lack of upward mobility. Some human resource managers reported that management positions still tend to have somewhat higher pay, but not enough to cause dissatisfaction among those who choose to advance in technical career paths. Successful use of dual career ladders demands careful attention to job evaluation procedures, and may require the nurturing of a climate that values technical contributions as well as management talent. Even this may be insufficient for firms that are no longer growing fast enough to accommodate mobility expectations. One large company has attempted to create a culture in which opportunity may be defined in terms other than promotion. An emphasis on challenge and development of new skills is likely to become increasingly important as more firms reach the end of their fast-growth stage.

Other retention strategies focus on the work itself and working conditions. One strategy is to hire recognized industry experts, to demonstrate a commitment to research and development. Our interviewees reported the importance of leading edge work with state of the art equipment. Other practices include maintaining an informal atmosphere with a lack of strict rules, and having dissatisfied people talk to others who have worked elsewhere under worse conditions to gain perspective on expectations.

Productivity

Most of the strategies used to improve productivity deal with providing rewards for desired performance. Group incentives as large as 20 percent of salary are used to motivate people to work long hours to meet deadlines, providing a competitive advantage in faster product development. A survey of human resource managers conducted by Miljus and Smith (1987, p. 127) revealed that in fast-growing organizations, employees were rewarded for high performance with advancement and merit compensation: "These organizations expected and received high quality performance. Employees believed that they were the most important asset and that they were respected by management. Reasonable risk taking and learning were encouraged." The financial performance of such organizations benefitted from better productivity and responsiveness to market opportunities. Some of our interviewees described efforts to create an atmosphere conducive to innovation and high performance, such as open communication (open door policies and skip level conferences), clear job expectations and an emphasis on feedback.

Emphasis on careful selection to assure productivity also was reported. Selection interviews included sample problems to assess technical knowledge and the ability to use it. One interviewee commented that he prefers hiring a "superstar" who might stay only three or four years to hiring a more average person who might stay ten years or longer.

SUGGESTED PRACTICES

Human resource managers face immediate problems with attraction and retention, so it seems necessary to make some preliminary suggestions of strategies to consider. Many practices that have been adopted seem appropriate for the conditions in which they were developed, although their effectiveness in achieving desired results rarely has been measured. Implementation of strategies patterned after those described above might be helpful if proper judgment is exercised as to their appropriateness for a particular situation. In addition, we can add some guidelines supported by research in other areas of management.

Attraction

It is difficult to offer many new strategies to attract scientists and engineers in the present labor market. The supply and demand situation requires attractive compensation packages and adequate relocation benefits. A company without sufficient capital to offer salaries at or above the market needs to compensate with bonuses, incentives or ownership shares of the company. Beyond that, a company relies on its reputation for providing interesting and stable employment in a satisfying work environment.

Retention

Reviews of the turnover process and retention strategies consistently document the importance of positive job attitudes, particularly satisfaction with the type of work (Hulin, in press; McEvoy & Cascio, 1985; Scott & Taylor, 1985). Sherman (1986) cited studies that showed autonomy to be negatively correlated with turnover and positively correlated with innovation. These findings indicate the importance of interesting and rewarding work, task significance, responsibility and the opportunity to use abilities to the fullest. Providing a variety of challenging assignments also may be a key to preventing obsolescence (Kleingartner & Mason, 1986). Therefore, McEvoy and Cascio's (1985) recommendation to consider job enrichment as a retention tool may be particularly germane to high technology firms.

Other elements of the high technology environment that have been reported to create dissatisfaction also should be addressed (Parden, 1981; Sherman, 1986). Opportunities for training and development to stay in touch with the state of the art and for advancement in technical career paths should be provided. Salaries should be reviewed periodically, or whenever someone leaves to accept a higher paying job, with special attention to alleviating salary compression.

Orientation and realistic job previews, especially with regard to management and administrative duties, could reduce unpleasant surprises and consequent dissatisfaction. As one personnel manager noted, there is a temptation to put urgently needed applicants to work immediately, before they have had an opportunity to become acclimated. Corning Glass Works found that an improved orientation program saved over $250,000 per year in turnover costs among professional employees (McGarrell, 1984), a finding consistent with McEvoy and Cascio's (1985) conclusion that realistic job previews reliably reduce turnover.

Many recommendations are possible for improving the management skills of scientists and engineers who supervise others. A fundamental problem is most engineering schools' lack of training in human relations skills, oral presentation and team projects. Because of this, training in supervisory skills, communication and team leadership is widely needed. McCarthy et al. (1987) noted that although technical skills often are a prerequisite, the successful R&D manager needs to supplement these with broad knowledge of the market and good interpersonal skills. Selecting for management ability and interest should be considered. Recruiting from the military can provide experienced engineers with extensive management training.

An alternative to changing conditions that lead to turnover is developing selection criteria that favor people who are less likely to quit (although that might conflict with selecting for highest performance). Weighted application blanks, for example, have been shown to predict tenure (Owens, 1976). Another option is to use human resource information systems to anticipate turnover of key personnel so that backups can be planned in advance, or to emphasize cross training to cope more easily with turnover.

Before leaving the topic of retention, it bears emphasizing that turnover need not be dysfunctional, especially if it is the poorer performers who are leaving. Performance appraisals should be used to determine who are the best people so special efforts can be made to retain them. Replacing those who quit or are terminated provides an important source of innovation (Allen, 1971). One firm we interviewed has a policy of maintaining a flow of new hires even during periods of decreased demand in order to enhance creativity. A similar dynamic was reported by Wells and Pelz (1966), who found scientific contributions of R&D work groups peaked at four to five years. Perhaps group membership should be rotated after such an interval.

Productivity

A general guideline for productivity is to provide meaningful rewards that are contingent upon desired performance. This is an old formula that works when it is implemented properly and consistently. Besides financial rewards, recognition of employees' achievements and their value to the organization is an effective reward. Gomez-Mejia and Balkin (1987) demonstrated the importance of reward contingencies. Profit sharing and stock options were correlated with reduced withdrawal cognitions, while performance-contingent bonuses were correlated with increased performance and satisfaction with pay, in addition to reduced withdrawal cognitions. Scientists and engineers work almost exclusively on team projects, so individual contributions are hard to measure, especially in short time frames (Katz & Allen, 1985). Thus, feedback and rewards probably should be group oriented.

A review of studies of organizational effectiveness (Steers, 1977) suggested that highly effective organizations are both achievement oriented and employee centered. This is probably true for high technology organizations as well as for others. Achievement orientation might include such practices as making sure priorities and standards are appropriate to the organization's needs and understood and agreed on by all. Also, the content of jobs should be reviewed to make sure technical professionals are able to commit their time efficiently and according to those priorities.

RESEARCH AGENDA

High technology industry is relatively young, and consequently, human resource management is being developed in response to the exigencies of current situations in particular organizations. As such, current strategies to attract and retain scientists and engineers have benefitted more from the experience and judgment of managers dealing with these issues in their daily activities than from scientific research. The effectiveness and generalizability of those strategies remain largely untested. Anderson and Kleingartner (1987) contended that high technology firms are especially vulnerable to piecemeal adoption of human resource practices that have been socially legitimized, but are not necessarily suited to the firm's needs at that time. It is also uncertain how currently popular strategies will need to be changed as companies evolve. The anecdotal nature of much of the existing data, and the resulting tentativeness of our recommendations speak to the need for research to measure the impact of current practices and to test recommendations for improving them.

Needed information falls into three major categories. The first concerns the nature of high technology employers. There still is much disagreement about exactly what constitutes high technology firms, and how they are unique from other employers. More sophisticated taxonomies are called for that incorporate product life cycle and business unit size. Balkin and Gomez-Mejia (1987) have shown the importance of these variables for determining the effectiveness of compensation strategies; similar insights into staffing and recruitment issues have been suggested by Kochan and Chalykoff (1987).

Another category is the characteristics of high technology employees. Successfully matching employee abilities and desires with employer needs and opportunities (Dawis & Lofquist, 1984) results in improved motivation and retention. Unfortunately, our current understanding of high technology workers is extremely limited. Articles imply that high technology workers are a special breed with unique needs, who create distinct demands on the organization, yet very few empirical data have been collected. Such basic strategies as attitude surveys would be very helpful, especially if conducted on databases large enough to allow generalizations across different local labor markets and types of organizations. Focused studies of the factors leading to a decision to join (or quit) a company also would be helpful. Moreover, future studies should direct such questions to employees or applicants rather than to managers or human resource professionals, whose responses may tell us more about managers' perceptions than workers' attitudes.

The third category includes research into the effectiveness of alternative attraction and retention strategies. While this issue is the one most pertinent to the present paper, we believe that it will be addressed most effectively only after we learn more about high technology firms and workers. As an example, it has been suggested that much mobility is due to young, "fast-tracking" professionals who seek to make their fortunes with equity in high-risk startup companies. Such individuals are not likely to be attracted to or remain with larger, more stable firms regardless of what policies are implemented. Yet very few existing data address the frequency with which either type of individual (or firm) is found, or the success of different strategies for matching such individuals and firms. If hiring and keeping workers is largely a function of a successful match between applicant and employer, such information is crucial to the development of successful attraction and retention programs.

In the meantime, better defining the nature of the problem will be useful. Assuming that high technology firms currently face personnel shortages, it is crucial that we determine how much this is due to attrition of current employees, expanded needs due to growth or simply inadequate supply to meet steady-state conditions. Reported turnover rates among high technology firms vary dramatically, but are seldom greatly in excess of typical rates. (The Bureau of National Affairs (1987) reports that turnover in 1986 averaged 12 percent, but was as high as 18 percent in the finance industry.) Turnover rates also need to be evaluated in the context of both local labor markets and overall economic conditions. For example, much of the highly publicized mobility of Silicon Valley employees occurred in the late 1970s, a period in which turnover rates for all sectors of the economy reached historical highs (22.8% in 1979). Getting reliable and up-to-date data concerning turnover rates and selection ratios is an obvious step toward determining whether and how large a problem exists.

Finally, we feel it is important to consider attraction and retention programs in a broader, system context. For example, before deciding to hire additional personnel, it may be worth asking whether efficiency of the current work force is adequate. This also suggests the question of tradeoffs between hiring for productivity versus longevity; it may be that the two goals are at least partially incompatible. At the least, each firm needs to determine for itself whether the problem (if it exists at all) is due to low productivity, excessive turnover, inadequate supply or an inability to attract qualified applicants.

It seems apparent that what we know about attraction and retention of scientists and engineers is overshadowed by what we only think we know and by what we do not know. The importance of filling these knowledge gaps is highlighted by Schmidt, Hunter and Pearlman's (1982) research on the economic payoff of hiring the most qualified workers, by the work of Boudreau and Berger (1985) on the costs of losing valued employees, and by demographic analyses illustrating the increasing importance of "knowledge workers" to the U.S. economy. We hope this paper will encourage researchers as well as managers to develop more systematic knowledge to guide theory and practice.

ACKNOWLEDGMENT

An earlier version of this paper was presented at the conference, Managing the High Technology Firm, University of Colorado at Boulder, January 14, 1988.

REFERENCES

Allen, T.J. (1977). *Managing the flow of technology*. Cambridge, MA: MIT Press.

Anderson, C.A., & Kleingartner, A. (1987). Human resource management in high technology firms and the impact of professionalism. In A. Kleingartner & C. A. Anderson (Eds.), *Human resource management in high technology firms* (pp. 7-35). Lexington, MA: Lexington Books.

Balkin, D.B., & Gomez-Mejia, L.R. (1985). Compensation practices in high technology industries. *Personnel Administrator, 30*(6), 111-123.

________(1987). Toward a contingency theory of compensation strategy. *Strategic Management Journal, 8,* 169-182.

Boudreau, J., & Berger, C. (1985). Decision-theoretic utility analysis applied to employee separations and acquisitions [Monograph]. *Journal of Applied Psychology, 70*(3), 581-612.

Bureau of National Affairs. (1987). Median job absence and turnover rates. *Policy and Practice Series: Personnel Management, 267,* 49-68.

Business woos engineers. (1979). *Dun's Review. 13*(6), 82-84.

Dawis, R., & Lofquist, L. (1984). *A psychological theory of work adjustment.* Minneapolis, MN: University of Minnesota Press.

Golson, H.L. (1985). The technically oriented personality in management. *IEEE Transactions on Engineering Management, EM-32,* 33-36.

Gomez-Mejia, L.R., & Balkin, D.B. (1987). *The effectiveness of individual and aggregate compensation strategies in an R&D setting.* Unpublished manuscript, College of Business and Administration, University of Colorado at Boulder.

Greenwald, H.P. (1978). Scientists and the need to manage. *Industrial Relations, 17,* 156-167.

Grissom, G.R., & Lombardo, K.J. (1985). The role of the high-tech HR professional. *Personnel, 62*(6), 15-17.

Hulin, C. (In press). Adaptation, persistence and commitment in organizations. In M. Dunnette (Ed.), *Handbook of industrial and organizational psychology* (2nd ed.). Chicago: Rand McNally.

Kail, J.C. (1987). Compensating scientists and engineers. In D.B. Balkin & L.R. Gomez-Mejia (Eds.), *New perspectives on compensation.* (pp. 278-281). Englewood Cliffs, NJ: Prentice-Hall.

Katz, R., & Allen, T.J. (1985). Project performance and the locus of influence in the R&D matrix. *Academy of Management Journal, 28,* 122-126.

Kleingartner, A., & Mason, R.H. (1986). Management of creative professionals in high technology firms. *Industrial Relations Research Association Proceedings,* April: 508-515.

Kochan, T.A., & Chalykoff, J.B. (1987). Human resource management and business life cycles: Some preliminary propositions. In A. Kleingartner & C.A. Anderson (Eds.), *Human resource management in high technology firms* (pp. 183-200). Lexington, MA: Lexington Books.

McCarthy, D.J., Spital, F.C., & Lauenstein, M.C. (1987). Managing growth at high technology companies: A view from the top. *Academy of Management Executive, 1,* 313-322.

McEvoy, G.M., & Cascio, W.F. (1985). Strategies for reducing employee turnover: A meta-analysis. *Journal of Applied Psychology, 70,* 342-353.

McGarrell, E.J. Jr. (1984). An orientation system that builds productivity. *Personnel Administrator, 29,* 75-85.

Miljus, R.C., & Smith, R.L. (1987). Key human resource issues for management in high tech firms. In A. Kleingartner & C.A. Anderson (Eds.), *Human resource management in high technology firms,* (pp. 115-131). Lexington, MA: Lexington Books.

Milkovich, G.T. (1987). Compensation systems in high technology companies. In D.B. Balkin & L.R. Gomez-Mejia (Eds.), *New perspectives on compensation* (pp. 269-277). Englewood Cliffs, NJ: Prentice-Hall.

Owens, W.A. (1976). Background data. In M. Dunnette (Ed.), *Handbook of industrial and organizational psychology,* (pp. 609-644). Chicago: Rand McNally.

Parden, R.J. (1981). The manager's role and the high mobility of technical specialists in the Santa Clara Valley. *IEEE Transactions on Engineering Management, EM-28*(1), 2-8.

Schmidt, F., Hunter, J., & Pearlman, K. (1982). Assessing the economic impact of personnel programs on work-force productivity. *Personnel Psychology,* 35, 333-347.

Scott, K., & Taylor, G. (1985). An examination of conflicting findings on the relationship between job satisfaction and absenteeism: A meta-analysis. *Academy of Management Journal, 28,* 599-612.

Seelig, P. (1985). Where it all begins. *Incentive Marketing, 159*(4), 68-72.

Sherman, J.D. (1986). The relationship between factors in the work environment and turnover propensities among engineering and technical support personnel. *IEEE Transactions on Engineering Management, EM-33*(2), 72-78.

Steers, R.M. (1977). *Organizational effectiveness: A behavioral view.* Santa Monica, CA: Goodyear.

Wanted: High-tech engineers. (1985). *Dun's Business Month. 125*(3), 35-36.

Wells, W., & Pelz, D. (1966). Groups. In D. Pelz & F. Andrews (Eds.), *Scientists in organizations: Productive climates for research and development* (pp. 79-103). New York: Wiley.

PERFORMANCE EVALUATION IN HIGH TECHNOLOGY FIRMS:
PROCESS AND POLITICS

Gerald R. Ferris and M. Ronald Buckley

INTRODUCTION

High technology firms are attracting more attention from organizational scientists and practitioners than perhaps any other industrial sector, primarily because of the particular managerial challenges that have emerged in such environments. A major set of these challenges are concerned with the effective management of human resources. One of the most central human resources activities, which can be effectively integrated with many others, is performance evaluation. Performance evaluation systems, when properly implemented, can make a major contribution to the productivity and overall effectiveness of high technology firms, and the effective design and implementation of such systems for Research and Development (R&D) professionals continue to pose a major challenge. Despite the fact that high technology firms consistently have been presented as exemplary organizations for which to work (e.g., Levering, Moskowitz, & Katz, 1984), relatively little is known about the human resources practices that have facilitated this image. The present paper is part of a continuing effort to enhance our understanding in this area through a survey of characteristics of performance evaluation systems in R&D units (i.e., for scientists and engineers) of high technology firms.

High Technology Environments

The environments of high technology firms tend to be characterized by volatile markets and rapidly changing technologies, which result in a high degree of ambiguity and uncertainty (e.g., McCarthy, Spital, & Lauenstein, 1987). McCarthy et al. noted that for firms to operate effectively in the high technology environment, they must be capable of "periodic shifts between chaos and continuity," they must be able to make timely decisions under conditions of uncertainty, and, by implication, maintain a high degree of flexibility. The high technology environment thus seems to create a fundamental dilemma for organizational and management practices by the simultaneous need for both structure and flexibility.

The conditions created by characteristics of the external environment are intensified by the dynamics of the internal work environments of high technology firms. Of particular concern are the nature of specific occupational categories and of formal policy and procedure. Although a large percentage of jobs in high technology firms are classified as blue-collar, production-type in nature (Belous, 1987), an area of particular interest because of their unique characteristics and the special issues and challenges they pose is the professional occupational group (Anderson & Kleingartner, 1987; Bailyn, 1985; Raelin, 1985). One prominent issue that has emerged with respect to the management of professionals is the value they place on autonomy and their resultant dissatisfaction with and disregard for rules, policies, and so forth. Raelin suggested that professionals generally react negatively to a preponderance of rules that define actions. More specifically, in a study of an industrial research laboratory, LaPorte (1965) reported that scientists resisted all forms of rules or procedures (including "personnel evaluation"), and regarded such activities as sources of strain between themselves and management.

One might suggest that both the nature of the high technology work environment and the characteristics of professionals and their reaction to procedure would contribute to a general lack of formalization, with specific reference to a lack of formal personnel policy development to structure and systemize human resources decisions/actions. Recent survey results demonstrated that only one of the high technology firms surveyed reported having a formal management succession planning system (Peterson, 1985). Such findings suggest the need for a more systematic investigation of human resources management practices in high technology firms.

Human Resources Management Practices

It is evident that more attention is being devoted to human resources management in high technology firms (e.g., Kleingartner & Anderson, 1987).

The particular challenges presented by the dynamic environments of high technology firms and the characteristics of professional jobs (i.e., particularly scientists and engineers) have led organizational scholars to examine several key areas of human resources management. A number of authors have identified and discussed the attraction/staffing and retention of critical human resources as major challenges for high technology firms (Gomez-Mejia & Balkin, 1985; Grissom & Lombardo, 1985; Miljus & Smith, 1987). Related to these issues of attraction and retention, another focus of research attention has been compensation practices and the design of effective reward systems for professionals in high technology firms (Balkin & Gomez-Mejia, 1984, 1987; Milkovich, 1987; Von Glinow, 1985).

Other human resources management challenges posed by professional jobs in high technology firms today include training and development, career development and internal mobility (e.g., Bailyn, 1982; Dalton, Thompson, & Price, 1982; Ferris, 1988; Kanter, 1984). However, despite the increased attention to a number of important human resources management practices in high technology firms, a neglected but most critical area is performance evaluation.

Performance Evaluation Systems

Surprisingly little research attention has been devoted to the investigation of performance evaluation systems in high technology firms, despite the central role of such systems and their increased importance in various types of organizations today (e.g., Bernardin & Beatty, 1984). Deming suggested that performance evaluation represents the premier American management challenge (Peters, 1987), and Grove concluded, from his experience at Intel, that "giving performance reviews is a very complicated and difficult business and that we, as managers, don't do an especially good job at it" (1983, p. 182). Although the current status of knowledge in this area is admittedly limited, reflecting more practitioner-oriented, data-free testimonials and case studies and a lack of larger-scale empirical research, it does provide some useful information concerning what is and is not known about performance evaluation systems in high technology firms.

Criteria and Evaluation Methods

While there is some inconsistency in the recent literature, there appears to be considerable agreement concerning the use of performance objective based evaluation systems for professionals (Butler & Yorks, 1984; Cascio, 1986; Cox, 1982; Lawler, Mohrman, & Resnick, 1984; Mossholder & Dewhirst, 1980; Oliver, 1985; Oliver, Nussbaumer, & Grimmett, 1985; Peters, 1987; Smith & Tuttle, 1982). The inconsistency that has been observed

concerns the results reported by two recent surveys of performance evaluation practices. Laud (1984) surveyed the Fortune 1300 organizations and reported that, for professional (nonmanagerial) employees, the most frequently used performance evaluation systems involved the use of objectives-based and Management By Objectives (MBO), together account for 75% of the responses. Another survey, focusing on exempt professional-technical employees, conducted by the Bureau of National Affairs (1983), reported that the essay method was used most frequently (60% of the cases), followed by graphic rating scales (55%) and MBO (44%).

Another issue that has emerged with respect to methods of performance evaluation for professionals concerns the use of peer evaluations, and several investigators have discussed the applicability of this method in such contexts (Mossholder & Dewhirst, 1980; Oliver, 1985; Oliver et al., 1985; Peters, 1987; Raelin, 1985). However, Cascio (1986) and Peters (1987), in particular, have argued against the use of peer rankings (i.e., ranking an employee's performance relative to coworkers) for professionals. They believe that such comparisons would result in a number of "below average" ratings and suggest that it is more beneficial to compare professionals to performance standards, not to each other.

Subordinate Involvement

Contemporary management philosophies reflecting the importance of quality of work life have translated into a greater degree of employee input, participation and involvement in a number of human resources practices (Lawler, 1986). This increased involvement of employees has also materialized as greater subordinate participation in the performance evaluation process, through supervisor-subordinate mutual goal setting and self-evaluation (e.g., Teel, 1978). The increased involvement of professional employees in the performance evaluation process has been examined by several organizational scientists, largely indicating evidence in support of such participation (Butler & Yorks, 1984; Cocheu, 1986; Foulkes, 1987; Lawler et al., 1984; Mossholder & Dewhirst, 1980; Peters, 1987). Interestingly, in his recent book, in which he devotes an entire chapter to performance evaluation, Andrew Grove (The President of Intel, "one of the nation's premier high technology companies") argued that the evaluation of performance is the job of the supervisor, and that self-review should not be used!

A survey of performance evaluation practices at four U.S. Navy R&D laboratories (with relevance for professionals in high technology firms) reported some concerns about the nature of evaluation for professionals (Nigro, 1981). In response to the question "My performance rating represents a fair and accurate picture of my actual performance," 57% of the respondents "disagreed" or "didn't know." Furthermore, 47% of the respondents indicated "disagree" or "didn't know," respectively, to the questions, "The standards used to evaluate my performance have been fair and objective," and "In the past I have been aware of what standards have been used to evaluate my performance." Such problem areas could reflect inadequate human resources policy development, but it might also suggest inadequate use of the performance evaluation system by supervisors. In the identification of key problem areas of performance evaluation (primarily focusing on high technology firms), it is consistently mentioned that supervisors inadequately use human resources systems and inconsistently integrate performance evaluation information used in important human resources decisions (Butler & Yorks, 1984; Cocheu, 1986; Oliver, 1985). Such results suggest the possibility that factors such as organizational politics have at least some influence upon the process and outcomes of performance evaluation for professionals in high technology firms.

Organizational Politics

In her typology of human resource cultures, focusing on the attraction and retention of professional employees, Von Glinow (1985) identified specific instances (e.g., "Apathetic Cultures") in which the evaluation of performance was based on politics. Frost (1987) has recently discussed the importance of organizational politics in human resources management, and Ferris and his colleagues have been involved in a program of research designed to better understand the dynamics of political behavior in organizations (e.g., Fandt & Ferris, 1990; Ferris, Fedor, Chachere, & Pondy, 1989; Ferris, Russ, & Fandt, 1989). Longenecker, Sims, and Gioia (1987) have argued, and provided some evidence to support the notion, that supervisors use performance appraisals politically in order to maximize their own self-interests or achieve personal agendas. One could reasonably argue that the special features of professionals' jobs and high technology work environments might provide an opportunity for organizational politics to influence important human resources decisions/actions, including performance evaluations. In fact, Ouchi (1980) argued that in new technology organizations, the rate of change and ambiguity of performance evaluation may simply overwhelm any attempts at rational control.

The foregoing review suggests a need for further investigation in order to gain more current, state-of-the-art knowledge concerning performance evaluation systems for professionals in high technology firms. The present study is an exploratory effort to address that need. The performance evaluation practices of a representative set of high technology firms were surveyed concerning system and process characteristics, major problem areas and the potential influence of organizational politics. The survey was confined to performance evaluation systems for R&D professionals (i.e., scientists and engineers).

METHOD

Sample

The sample of firms used in this study consisted of 104 high technology firms selected to insure variability on size and main product lines. Furthermore, because the focus of this investigation was on performance evaluation systems in R&D units of high technology firms, the organizations selected for inclusion had to have at least one R&D unit and there had to be a reasonable representation of scientists and engineers. Product lines for firms in the sample included electronics, technical instruments, pharmaceuticals, health-care products, medical instrumentation and biotechnology.

Surveys were mailed to the human resources departments (routed to the director) of the 104 firms with guarantees of strict confidentiality, and useable responses were returned by 32 of the firms. Male respondents made up 72% of the respondents, and mean age was 42.58 years, with average organization tenure of 8.15 years. 53% of the respondents were private, for profit organizations, and 34% were in the public sector. There was reasonable variability on organization size (i.e., number of employees) of respondents, with a mean of about 4000 employees. On average, respondents indicated that about 28% of their operation was R&D.

Survey

The survey used in this study was developed based on an extensive literature search as well as interviews with managers, scientists and engineers working in R&D units of organizations locally. The resulting instrument consisted of four sectors. The first section included questions relating to characteristics of the firm's performance evaluation system(s) for employees in their R&D units. The second section focused on process issues of evaluation, including the feedback process and the nature of supervisor-subordinate working relationships. Section three included questions concerning perceptions of the R&D work environment, particularly with respect to unfairness and political behavior. The final section included questions about organizational characteristics and background information.

RESULTS

Because this investigation was an exploratory effort to gain state-of-the-art information about R&D performance evaluation systems, and because the sample size is small, it seemed most appropriate to analyze the data in a largely qualitative manner. Frequencies and/or percentages were tabulated for the response categories under each item on the survey. Several open-ended questions were included in the survey and, in those cases, results of content analyses are reported.

Performance Evaluation System Characteristics

Of the 32 respondents to the survey, 27 reported that there was a formal performance appraisal system in place for scientists and/or engineers in the R&D function. Two organizations reported having no formal system, but both of those firms mentioned that they conducted informal appraisals referred to as "reviews of work."

Most of the respondents reported using performance evaluations on R&D employees for both administrative *and* developmental purposes ($n = 26$), whereas two firms indicated that appraisals were used only for developmental purposes, and one firm reported administrative use only. Also, the majority of respondents indicated that formal evaluations were conducted once a year ($n = 26$). Of the remaining firms, one reported a less than once a year frequency (this firm indicated that the longer the employee works for the organization, the less frequent formal performance evaluations they receive), three reported a twice a year frequency, and one reported that appraisals were done more than three times a year (this organization mentioned that quarterly, informal performance reviews, not concerned with salary, were conducted).

Of the 32 responding organizations, 23 reported that just one standardized performance evaluation system was used for all employees within the R&D function, whereas six firms indicated that more than one system was used (i.e., different systems for different occupational groups). This could be a potential problem source if, in fact, there are different occupational categories represented in the R&D unit, and a single performance appraisal instrument or system is being force-fit to all. In fact, doing this would presumably minimize any benefits that could accrue from participation in the developmental phase of performance evaluation. Participation in this process is more likely to result in the development of a more appropriate/relevant performance evaluation instrument and a closer relationship between scientific activities and organizational goals. Furthermore, such an approach should contribute to more accurate perceptions of the jobs of scientists and engineers by their supervisors.

Surprisingly, a large majority (23 of 32 responding firms) of the respondents reported that those responsible for managing/supervising scientists and engineers in the R&D function had formal technical training in the areas in which they evaluated. This tends to debunk the prevailing myth that R&D managers have minimal technical training in the areas in which they supervise. An interesting twist to this, however, comes from a somewhat dated research report. Hirsch, Milwitt and Oakes (1958) documented that the average professional (scientist) reported spending about two-thirds of his or her time on routine work that did not require a professional-technical background. Thus, if this finding is still valid today, a supervisor would not need an extensive technical background to adequately evaluate such performance.

Sources of evaluation outside of the supervisor-subordinate dyad apparently have little input into the performance appraisal process. All of the responding firms reported using the immediate supervisor as a source of evaluation, and approximately one-third of the firms used subordinate self ratings of performance in conjunction with information from the immediate supervisor. This seems to be consistent with the notion that supervisors have adequate technical background to effectively conduct performance evaluations, thus, they might not need to seek additional information from alternative sources. Two responding firms reported using peers within the R&D function as sources of appraisal, and two firms indicated that they used outside (of the organization) professionals in the employee's field of expertise. The lack of reliance on peers as a source of appraisal for scientists and/or engineers in R&D units seems surprising in light of the literature, which has tended to argue in favor of such an evaluation source for these types of employees (e.g., Oliver et al., 1985; Raelin, 1985).

The type of instrument, scale or method most frequently used in performance evaluation systems of responding organizations was some variant of goal setting or Management by Objectives (MBO) (24 of 32 respondents; 75%). Fewer respondents reported using graphic rating scales (n = 12) and behaviorally-anchored rating scales (BARS) (n = 9). One responding firm reported using an essay-type performance appraisal instrument. These results are substantially different from those reported in a fairly recent survey of performance appraisal techniques used with exempt professional-technical employees (BNA, 1983). Those results reported that the essay method was used in about 60% of the cases, graphic rating scales in about 55% of the cases, MBO (44%), checklists (28%), critical incidents like BARS (28%), and straight ranking (8%). Like the present results concerning the instruments or methods used, these results add up to more than 100% because some organizations reported using more than one performance evaluation method.

An open-ended question was included, which asked respondents to report the performance criteria or dimensions on which R&D professionals were evaluated, and how such criteria were derived. The results of a content analysis of responses to this question are presented in Table 1. As can be

Table 1. Performance Criteria for R&D Professionals and Their Derivation

Criteria	*Frequency*
Meeting Objectives	
In cost-effective manner	5
In timely manner	6
Actual results (also mentioned as progress achieved or quantity of work produced)	13
Initiative	4
Creative/innovative	7
Quality of work	7
Number of successes relative to failures	1
Personal Skills	
Intrapersonal and interpersonal skills	4
Client relations	2
Teamwork/communication skills	5
Dependability	2
Aggressiveness	1
Work Attitude	2
Technical/Professional Skills	
Technical/professional know-how	7
"State-of-the-art" knowledge	1
Professional growth	2
Administrative Skills	
Selling/promotion	1
Management skills	6
Judgement/problem-solving skills	4
Operating methods	3
Leadership	3
Standard performance appraisal forms (based on supervisors' judgement)	1
Peer feedback and performance against objectives	2
Goals and objectives established by supervisor and employee for review purpose	8
Progress against MBO	2
Peer vs. peer rating based on quality/quantity	1
All ranked within group level annually against established criteria	1
Criteria different for each person	1

seen, four broad criterion dimensions were identified, through content analysis, with the frequency of specific responses under each. Overall, when asked about criteria, the most often mentioned was actual work accomplished ($n = 13$). This was either mentioned as progress achieved, objectives met, projects completed or quantity of work done. Quality of work, professional/technical knowledge, and creativity were mentioned second most frequently (i.e., 7 times each), and management skills was third in frequency ($n = 6$).

Respondents reported several ways in which criterion dimensions were derived. By far, the most frequently mentioned was reviewing the employee against goals and objectives established by both the supervisor and the employee. This is consistent with the reported widespread use of goal setting/MBO methods of performance evaluation used with R&D professionals.

Characteristics of the Performance Evaluation Process

All of the respondents reported that they used some type of performance evaluation interview to inform the ratee of his or her performance ratings, although the timing of the interview and the filing of the formal appraisal were somewhat different. Six firms (out of 32) indicated that the immediate supervisor completes a performance evaluation form for each of his or her employees and a copy goes on file after the supervisor schedules an interview to tell the subordinate how he or she was evaluated. Seventeen of the respondents reported that the supervisor makes some notes on a working copy of a performance evaluation form for each of his or her employees, has an interview with each where open discussion of the person's performance is encouraged, and a final copy of the performance evaluation form is completed and signed by both parties and goes on file, which reflects the input and agreement of both. Seven responding firms reported "other" ways that the performance evaluation process operates in their organizations, including the employee completing a self-evaluation of performance and comparing it with supervisor's evaluation, using peer ranking and verbal work review.

Generally, respondents reported that performance-related feedback was provided on a reasonably timely basis to R&D professionals by their supervisors. Twenty-nine of the 32 firms reported this to be the case either sometimes, (n = 8), frequently (n = 16), or always (n = 5), whereas only one firm indicated that this was rarely done.

The majority of responding firms felt that the performance evaluations given to R&D professionals by their supervisors were accurate (n = 16 out of 32) or extremely accurate (n = 4). Nine firms rated the appraisals as somewhat accurate, and there were no indications of perceived inaccuracy.

The perceived helpfulness of performance appraisals was high, with 84% of the respondents perceiving appraisals to be at least somewhat or more helpful. A possible confound in the interpretation of this finding may be due to overconfidence bias; that is, if one does it or is responsible for it, it must be well done and helpful (e.g., Nisbett & Ross, 1980). Nigro (1981) reported that 57% of his respondents (i.e., scientists and engineers at four Navy R&D Laboratories) disagreed with or didn't know about the statement "My performance rating represents a fair and accurate picture of my actual job performance." Furthermore, Nigro presented results which demonstrate that performance evaluations were perceived as not being helpful in assessing one's strengths and weaknesses (47% of the respondents in R&D units were not fully aware of what standards had been used to evaluate their performance).

All respondents felt that supervisors kept informed concerning the work of their subordinates, although they displayed different degrees of awareness. Specifically, no one believed that supervisors remained far removed from their employees, leaving them to work autonomously with almost no feedback or guidance. The responding firms were divided nearly equally between perceiving that supervisors were aware of the employees' work progress and outcomes and provided some feedback and helpful suggestions (n = 13 out of 32), and reporting that the supervisor was actively aware of and involved in the employees' job and career progress, providing guidance and development in a mentoring sort of capacity (n = 14 out of 32).

According to the respondents, performance evaluations serve as a basis for a number of important human resources decisions/actions made about professionals in the R&D function. The frequencies of responses in each of the categories are presented in Table 2. It is clear, from an examination of the responses in Table 2, that performance evaluations are believed to play a more important role in some human resources decisions/action than in others. For example, with all firms responding in either the "Frequently" or "Always" response categories, performance evaluation seems to be seen as an important contributor to pay increases. Promotions also are perceived to be considerably influenced by performance appraisals, yet four of the 32 respondents (13%) believed that was only sometimes the case (see later discussion of organizational politics). For some of the other human resources decisions/actions, such as layoffs, bonuses and human resources and succession planning, there seems to be much less consistency in opinion. The responses concerning bonuses seem particularly surprising, with eight respondents reporting that performance appraisals rarely or never are used to make such determinations.

The next issue examined seems to create some inconsistency, because 56% of the respondents reported that there had been complaints from R&D professionals concerning the performance evaluation system or decisions/actions taken based on that system (10 firms indicated that there were rarely complaints and only one firm reported never having a complaint). This findings seems a bit contrary to respondents' perceptions of the effectiveness of the performance evaluation system, which suggested a high degree of accuracy and helpfulness. One possibility is that the systems actually are effective, but professional employees tend to complain a lot. Alternatively, the systems may not be as well developed as reported. Furthermore, although there have been complaints against the performance evaluation system, only two such cases were ever challenged legally.

Table 2. Use of Performance Appraisals in Human Resources Decisions/Actions

	Response Categories				
Decision/Action	*1* *Never*	*2* *Rarely*	*3* *Sometimes*	*4* *Frequently*	*5* *Always*
Pay increases	0	0	0	11	18
Promotions	0	0	4	16	9
Transfers	0	0	12	13	2
Terminations	0	2	4	8	14
Layoffs	2	4	5	7	8
Bonuses	3	5	1	7	8
Career planning	0	3	6	15	4
Performance counseling	0	0	4	17	7
Human resources planning	3	4	7	10	3
Succession planning	3	4	6	8	7

Note: Because of missing data, the sums of responses across categories for each decision/action are not the same.

Respondents were asked what were the major issues, problems and/or concerns with respect to the performance evaluation system for R&D professionals. The responses reported and their frequencies are presented in Table 3. A content analysis revealed essentially two broad areas of concern. One area is a group of systemic problems reflected by no direct link between performance and rewards, lack of accuracy in performance measurement, failure to define performance standards and constantly changing objectives. The other area seems to suggest some interpersonal concerns primarily represented by the belief that managers/supervisors do not use the system objectively, but also including a lack of participation in the process by subordinates, and a lack of dialogue between supervisors and subordinates.

Respondents also were asked for their suggestions for changing the performance evaluation system, and they offered a number of ideas, which are listed in Table 4. One would hope to find a close correspondence between areas of concern and suggestions for change, and in fact that does seem to be reflected from an examination of Tables 3 and 4. Suggestions for change range from more involvement/participation of subordinates in the appraisal process to the frequency of formal evaluation sessions (semi-annual and quarterly reviews were both suggested). Two themes of importance seem to emerge from the suggestions provided by respondents. One seems to be the need to objectify the situation (i.e., standards of performance, goals, rating, and so on). The other seems to focus on the behavior of managers/supervisors conducting the appraisals (i.e., behaving more objectively, using

Table 3. Major Issues, Problems and Concerns with Performance Appraisal System

Issue/Problem/Concern	*Frequency*
Difficulty in defining goals	1
Difficulty in defining work	1
Performance not tied directly to rewards	1
No accurate performance-related system	2
Managers/supervisors that don't use the system objectively	6
Job/objectives difficult to measure quantitatively	4
Objectives constantly change	1
Need more dialogue between employee and supervisor	2
Appraisal isn't tied enough to job description	1
Balance fostering innovation vs. MBO (timing)	2
Failure to define standards of performance	2
Consistency	1
Communicating adverse information	1
Employee does not participate enough in appraisal	1
System should be more developmentally or future oriented	2
No problems	3

Table 4. Suggestions for Changing the Performance Appraisal System

Suggestion	*Frequency*
Program based more closely on objectives and documented results	1
Measure performance directly against specific objectives/duties	1
Require direct employee input, involvement, self-evaluation	2
Promote discussion on employees' strengths and improvement areas	1
Set specific goals for next period	
Have required quarterly reviews	1
Hold semi-annual reviews	1
Hold supervisors accountable for development	1
Eliminate standard forms	1
Eliminate ranking	1
Better definition of how performance is measured	1
Establish standards on departmental basis	1
Require higher level management review	1
Use same system for the whole company, not just R&D	2
Separate developments from pay on performance aspects	1
Train supervisors on how to develop career goals	1
More time/money explaining review process and system to employees	1
Remove number scale, make appraisal more development oriented	1
Have subordinates prepare objective/development needs with management agreement	1
Tie performance to pay	1
Have manager/supervisor use system correctly	1

system correctly, and so on). Overall, these results seem to highlight two major areas of concern, and areas in need of change that are actually related: (1) managers/supervisors not using the system objectively; and, (2) difficulty in quantitatively measuring standards and objectives of work performance. One could reasonably argue that both the nature of the work environments of high technology firms and the characteristics of the jobs of scientists and engineers working in R&D units in organizations contribute to a high level of uncertainty and ambiguity, and preclude the type of objectivity typically sought. This has been discussed briefly in an earlier part of this paper, and it will be further examined in the last section. Alternatively, one might suggest that such uncertain and ambiguous work environments provide fertile ground for the emergence of organizational politics to influence decisions and actions. This issue is examined next.

Organizational Politics

Respondents were asked to report the extent to which they believed that organizational politics influenced a number of important human resources decisions/actions. The frequencies of responses in each of the response categories is presented in Table 5. From an examination of Table 5, the majority of respondents seem to believe that politics have a "sometime to frequent" effect upon promotions and transfers and pay increases. (This seems consistent with the finding reported earlier that 13% of the respondents believed that only sometimes were performance appraisals used in making promotion decisions.) For performance appraisals, 50% of the responding firms reported that politics rarely was involved in decisions/actions, whereas 41% indicated that it was either "sometimes" or "frequently" an influence. For disciplinary actions, terminations, and layoffs, there appeared to be more evidence that politics does not play a major role, although for both layoffs and for bonus decision, 19% of the respondents felt that there was some evidence of political influences.

In addition to their opinions concerning the influence of organizational politics in specific human resources decisions/actions, respondents were also asked to report the extent to which supervisors are "brutally frank" concerning the levels of performance. While there appeared to be considerable variability in responses, most respondents reported that supervisors were not "brutally frank" about the performance levels of subordinates (13 out of 32 believed they were not, 8 believed they were and 8 were undecided). These responses could be interpreted in several ways. One interpretation could be that this reflects a leniency bias in both performance ratings and feedback. Bass (1956) suggested that there are more compelling reasons to *not* be accurate (or "frank") in performance ratings,

Table 5. Influence of Organizational Politics on Resources Decisions/ Actions

	Response Categories				
Decision/Action	*1* *Never*	*2* *Rarely*	*3* *Sometimes*	*4* *Frequently*	*5* *Always*
Performance appraisal	0	16	11	2	0
Promotions and transfers	0	10	10	9	0
Pay increases	2	12	12	3	0
Disciplinary actions	4	19	5	1	0
Terminations	7	14	7	1	0
Layoffs	10	10	4	2	0
Bonuses	7	11	3	3	0

Note: Because of missing data, the sums of responses across categories for each decision/action are not the same.

relative to supervisory self-serving strategies. Longenecker et al. (1987) recently have suggested that supervisors use performance appraisals politically to maximize their own self interests or particular agendas, which may involve either inflation or deflation of evaluations.

Overall, the pattern of responses seem to suggest that organizational politics play more than a casual role in several important human resources decisions/actions. One is typically led to believe that such important decision/actions involve the systematic treatment of individuals based upon job-relevant information only. One reaction is that such a belief is simply naive. However, a more balanced perspective might suggest that different types of environments or contests allow for (or even encourage) more political behavior than others (Ferris et al., 1989). Politics tends to thrive on uncertainty, and both the nature of the work environments of high technology firms and the characteristics of the jobs of R&D professionals (i.e., scientists and engineers) contribute to a high level of uncertainty or ambiguity, and to subjectivity. Thus, it seems quite likely to find evidence of politics influencing decisions and actions taken in such contexts. Why political behavior did not appear to be even more prevalent than the levels reported perhaps reflects reality, or it might reflect bias on the part of respondents toward not admitting to engaging in such types of behavior (or being associated with it in any way). The study of "sensitive topics" such as organizational politics presents considerable difficulty (Frost, 1987). So, while the evidence of politics reported in the present study might not appear to be particularly strong, social desirability bias might suggest that it potentially reflects a quite conservative estimate of the true conditions.

DISCUSSION

The results of the present survey of performance evaluation systems for R&D professionals in high technology firms are interesting in light of previous reports. The responses to questions concerning performance criteria, derivation of those criteria and evaluation methods suggest the use of an objectives-based or goal setting system focusing on actual results or progress made toward accomplishment of goals and objectives established by supervisor and subordinate. However, 38% of the respondents reported using graphic rating scales that focused on criteria such as work attitude, effort, aggressiveness, and promptness. Furthermore, concerning performance evaluation techniques, while suggestions have been made in favor of peer review, its use was found to be virtually nonexistent.

In reviewing the major problem areas identified, interestingly, the problem mentioned most frequently was managers/supervisors not using the performance evaluation system objectively. In his discussion of common performance evaluation problems, four of the eight problems identified by Oliver (1985) related to the supervisors inappropriate use of the system. In two studies of performance evaluation in high technology firms, both identified the negative view of, and/or low confidence in, the performance evaluation process by supervisors (Butler & Yorks, 1984; Cocheu, 1986). Nigro (1981) reported that 46% of the R&D professionals surveyed responded "disagree" or "don't know" to the statement, "My job performance is carefully evaluated by my supervisor." Grove (1983) reported that at Intel, managers did not do a particularly good job of performance evaluation. These results, in conjunction with the present findings, suggest that supervisors do not use the performance evaluation system objectively and use it perhaps even inappropriately. What is unclear is whether such inappropriate use is based on negligence or lack of attention, or whether it is based on more active efforts of supervisors to use the system for some political agenda (e.g., Longenecker et al., 1987).

Another important issue that emerged from an examination of performance evaluation problem areas seem to be the nonperformance-based nature of such systems. This finding, in conjunction with the considerable proportion of respondents indicating that trait-based scales (focusing on work effort and personal traits) were used as a basis of performance evaluation system, is interesting when considering previous research. Butler and Yorks (1984) cited measurement instruments that stress personal traits instead of job performance as a major reason for performance evaluation system failure. In her typology of human resource cultures for professionals, Von Glinow (1985) suggested that in cultures characterized by a strong concern for people but weak performance expectations (i.e., "Caring Culture"), performance evaluations tend not to be performance-oriented. Rather, nonperformance related criteria are used, like cooperation, teamwork and fitting in, and it is quite likely that individuals would be evaluated on the basis of effort instead of results. One might argue that the less objective the performance outcomes of a job, the less sensitive are performance evaluation systems in detecting differences in true or actual work performance. Such systems then would tend to focus on the detection of differences in perceived performance, which can be influenced by symbolic (or political) behavior/performance. It is not surprising that in such situations, people are frequently evaluated on the basis of work effort or attitude (Pfeffer, 1981). To the extent that goals set are a manifestation of one's effort or work attitude, the results of a study of Dossett and Greenberg (1981) are quite interesting and relevant. They found that regardless of whether workers succeeded or failed in meeting the goals they set, supervisors rated the performance of workers setting higher goals as more favorable than those setting lower goals.

These results seem to suggest that the potential exists for political influences in performance evaluation and other human resources decisions/actions. In fact, in two of the four cells of Von Glinow's (1985) typology, political issues emerge as significant sources of influence on performance evaluations. The results of the present study suggest some perceived influence of organizational politics primarily with respect to promotions and transfers, pay increases and performance evaluations. The finding that supervisors do not use the performance evaluation system objectively might add further definition to the present results concerning politics.

Considerable variability was noted in the responses to the present performance evaluation survey, concerning both practices and politics. Such variability potentially could be due to boundary conditions such as the life cycle and associated strategy of the firm (Galbraith & Nathanson, 1978), or the type of human resources culture of the particular organization (Von Glinow, 1985). Both potential influences suggest considerable impact on performance evaluation systems.

Several cautions need to be noted in the interpretation of the results of the present survey. First, because it is based on a small sample of high technology firms, (N = 32), caution should be exercised in the "over interpretation" of the results as necessarily generalizable to other firms in the industry. Second, this survey solicited the responses of human resources managers (i.e., typically the director), not actual R&D professionals. It is reasonable to expect that responses might differ considerably, reflecting quite different perceptions of performance evaluation systems, as a function of one's position in the organization. Lawler et al. (1984) recently provided evidence in support of this notion, demonstrating that manager and subordinate perceptions of the performance evaluation process differed somewhat—sometimes considerably.

A third caution concerns the interpretation of the organizational politics responses. The reactive nature of the politics area and the managerial positions of respondents, typically associated with human resources policies and actions, would suggest possible bias in responses. Organizational politics is a sensitive topic and a typical response would be to not associate oneself (or organization) with political influences. Thus, the potential for social desirability would suggest that the obtained results could reasonably be assumed to be a conservative, lower-bound estimate of actual events.

Future research should expand on the present investigation with larger samples and additional areas of examination. One area that seems particularly important to examine in future research, as it related to performance evaluation systems in high technology firms, is the career development and internal mobility of professionals. Miljus and Smith (1987) identified this as a key issue or challenge for high technology firms in the future, and recent research results suggest a lack of formal human resources policy development to structure decision making concerning internal mobility (Peterson, 1985). One issue in particular that might be examined is the usefulness of the "fast-track" approach versus a slower evaluation and promotion system. Ferris (1988) has suggested that one might reasonably question the "fast-track" system, relative to the potential organizational implications, for at least two reasons. First, rapid movement through a series of jobs does not permit enough time in grade to develop one's skills and competencies to their fullest. Furthermore, in the long-term, a sustained contribution being made by people in such a system seems highly questionable. Thus, they don't appear as valuable to the organization.

Second, by virtue of such a "fast-track" system involving quick movement, one could argue that it encourages at least as much symbolic behavior (perhaps political in nature) as actual effective performance. Because one is in a position a reasonably short period of time, and because standards of performance on such jobs are ambiguous at best, individuals are likely to be evaluated more on how much it appears that they are contributing than on the basis of their actual (objective) performance level (Pfeffer, 1981). A system encouraging slower evaluation and promotion should help to effectively address a growing concern of high technology firms, that managers need to develop broad sets of skills in order to circumvent obsolescence (Kanter, 1984).

Stata (1986) identified human resources as "the most fundamental and enduring limitations to growth" of high technology firms. We concur. These and other issues require further investigation in order to develop a more informed understanding of effective human resources management practices in high technology firms.

REFERENCES

Anderson, C.S., & Kleingartner, A. (1987). Human resource management in high technology firms and the impact of professionalism. In A. Kleingartner & C.S. Anderson (Eds.), *Human resource management in high technology firms* (pp. 3-21). Lexington, MA: Lexington Books.

________(1987). Toward a contingency theory of compensation strategy. *Strategic Management Journal, 8,* 169-182.

Balkin, D.B., & Gomez-Mejia, L.R. (1984). Determinants of R & D compensation strategies in the high tech industry. *Personnel Psychology, 37,* 635-650.

Bailyn, L. (1985). Autonomy in the industrial R&D lab. *Human Resource Management, 24,* 129-146.

Bass, B.M. (1956). Reducing leniency in merit ratings. *Personnel Psychology, 9,* 359-369.

Belous, R.S. (1987). High technology labor markets: Projections and policy implications. In A. Kleingartner & C.S. Anderson (Eds.), *Human resource management in high technology firms* (pp. 25-45). Lexington, MA: Lexington Books.

Bernardin, H.J., & Beatty, R.W. (1984). *Performance appraisal: Assessing human behavior at work.* Boston: Kent.

Bureau of National Affairs (1983, February). Performance appraisal programs. *Personnel Policies Forum* (Survey No. 135). Washington, DC: The Author.

Butler, R.J., & Yorks, L. (1984). A new appraisal system as organizational chante: GE's task force approach. *Personnel, 61,* 31-42.

Cascio, W.F. (1986). Technical & mechanical job performance appraisal. In R.A. Berk (Ed.), *Performance assessment: Methods & applications* (pp. 361-375). Baltimore, MD: The Johns Hopkins University Press.

Cederblom, D. (1982). The performance appraisal interview: A review, implications, and suggestions. *Academy of Management Review, 7,* 219-227.

Cocheu, T. (1986). Performance appraisal: A case in points. *Personnel Journal, 64,* 48-55.

Cox, A. (1982). *The Cox report on the American corporation.* New York: Delacorte Press.

Dalton, G.W., Thompson, P.H., & Price, R.L. (1982). The four stages of professional careers: A new look at performance of professionals. In R. Katz (Ed.), *Career issues in human resource management,* (pp. 129-153). Englewood Cliffs, NJ: Prentice-Hall.

Dossett, D.L., & Greenberg, C.I. (1981). Goal setting and performance evaluation: An attributional analysis. *Academy of Management Journal, 24,* 767-779.

Fandt, P.M., & Ferris, G.R. (1990). The management of information and impressions: When employees behave opportunistically. *Organizational Behavior and Human Decision Processes, 45,* 140-158.

Ferris, G.R. (1988). The "processing" of human resources in the high technology industry. In L.R. Gomez-Mejia (Ed.), *Human resource management strategies in the high technology industry.* Symposium presented at the Academy of Management, 48th Annual National Meeting, Anaheim, California.

Ferris, G.R., Fedor, D.B., Chachere, J.G. & Pondy, L.R. (1989). Myths and politics in organizational contexts. *Group & Organization Studies, 14,* 83-103.

Ferris, G.R., & Kacmar, K.M. (1988). *Organizational politics and affective reactions.* Paper presented at the 30th Annual Meeting of the Southwest Division, Academy of Management, San Antonio, Texas.

Ferris, G.R., Russ, G.S., & Fandt, P.M. (1989). Politics in organiztions. In R.A. Giacalone & P. Rosenfeld (Eds.), *Impression management in the organization* (pp. 143-170). Hillsdale, NJ: Erlbaum.

Foulkes, F. (1987). Human resources at Auto Tel, Inc. In A. Kleingartner & C.S. Anderson (Eds.), *Human resource management in high technology firms* (pp. 85-102). Lexington, MA: Lexington Books.

Frost, P.J. (1987). *Political influences on human resource management practice.* Distinguished speaker presentation at the International Personnel and Human Resource Management Conference, Singapore.

Galbraith, J.R., & Nathanson, D. (1978). *Strategy implementation: The role of structure and process.* St. Paul, MN: West.

Gomez-Mejia, L.R., & Balkin, D.B. (1985). Managing a high tech venture. *Personnel, 62,* 31-36.

Grissom, G.R., & Lombardo, K.J. (1985). The role of the high-tech human resources professional. *Personnel, 62,* 15-17.

Grove, A.S. (1983). *High output management.* New York: Random House.

Hirsch, I., Milwitt, W., & Oakes, W.J. (1958). Increasing the productivity of scientists. *Harvard Business Review, 36,* 67-70.

Kanter, R.M. (1984). Variations in managerial career structures in high technology firms: The impact of organizational characteristics on internal labor market patterns. In P. Osterman (Ed.), *Internal labor markets* (pp. 109-131). Cambridge, MA: MIT Press.

Kleingartner, A., & Anderson, C.S. (1987). *Human resource management in high technology firms.* Lexington, MA: Lexington Books.

Laud, R.L. (1984). Performance appraisal practices in the Fortune 1300. In C.J. Fombrun, N.M. Tichy, & M.A. Devanna (Eds.), *Strategic human resource management* (pp. 87-100). New York: Wiley.

Lawler, E.E., III (1986). *High involvement management.* San Francisco, CA: Jossey-Bass.

Lawler, E.E., Mohrman, A.M., Jr. & Resnick, S.M. (1984). Performance appraisal revisited. *Organizational Dynamics, 13,* 20-35.

Levering, R., Moskowitz, M., & Katz, M. (1984). *The 100 best companies to work for in America.* Reading, MA: Addison-Wesley.

Longenecker, C.O., Sims, H.P., Jr., Gioia, D.A. (1987). Behind the mask: The politics of employee appraisal. *Academy of Management Executive, 1,* 183-193.

McCarthy, D.J., Spital, F.C., & Lauenstein, M.C. (1987). Managing growth at high-technology companies: A view from the top. *Academy of Management Executive, 1,* 313-323.

Miljus, R.C., & Smith, R.L. (1987). Key human resource issues for management in high tech firms. In A. Kleingartner & C.S. Anderson (Eds.), *Human resource management in high technology firms* (pp. 115-131). Lexington, MA: Lexington Books.

Milkovich, G.T. (1987). Compensation systems in high technology companies. In A. Kleingartner & C.S. Anderson (Eds.), *Human resource management in high technology firms* (pp. 103-114). Lexington, MA: Lexington Books.

Mossholder, K.W., & Dewhirst, H.D. (1980). The appropriateness of management-by-objectives for development and research personnel. *Journal of Management, 6,* 145-156.

Nigro, L.G. (1981). Attitudes of federal employees toward performance appraisal and merit pay: Implications for CSRA implementation. *Public Administration Review,* Jan.-Feb., 84-86.

Nisbett, R., & Ross, L. (1980). *Human inference: Strategies and shortcomings of social judgment.* Englewood Cliffs, NJ: Prentice-Hall.

Oliver, J.E. (1985). Performance appraisals that fit. *Personnel Journal, 64,* 66-77.

Oliver, J.E., Nussbaumer, C., & Grimmett, D.R. (1985). Adapting performance appraisal systems to changed technologies. *Human Systems Management,* 323-331.

Ouchi, W.G. (1980). Markets, bureaucracies, and clans. *Administrative Science Quarterly, 25,* 129-141.

Peters, T. (1987). *Thriving on chaos: Handbook for a management revolution.* New York: Knopf.

Peterson, R. (1985). Latest trends in succession planning. *Personnel, 62,* 47-54.

Pfeffer, J. (1981). Management as symbolic action: The creation and maintenance of organizational paradigms. In L.L. Cummings & B.M. Staw (Eds.) *Research in organizational behavior* (Vol. 3, pp. 1-52). Greenwich, CT: JAI Press.

Raelin, J.A. (1985). The basis for the professional's resistance to managerial control. *Human Resources Management, 24,* 147-175.

Smith, H.L., & Tuttle, W.C. (1982). Managing research scientists: Problems, solutions, and an agenda for research. *Journal of the Society of Research Administrators, 13,* 31-38.

Stata, R. (1986). Human resources — the limitation to high technology growth. In F.K. Foulkes (Ed.), *Strategic human resources management: A guide for effective practice* (pp. 43-55). Englewood Cliffs, NJ: Prentice-Hall.

Teel, K.S. (1978). Self-appraisal revisited. *Personnel Journal, 57,* 364-367.

Von Glinow, M.A. (1985). Reward strategies for attracting, evaluating, and retraining professionals. *Human Resource Management, 24,* 191-206.

TECHNOLOGICAL ADOPTION AND ORGANIZATIONAL ADAPTATION: DEVELOPING A MODEL FOR HUMAN RESOURCE MANAGEMENT IN AN INTERNATIONAL BUSINESS ENVIRONMENT

Urs E. Gattiker

Why should one specifically use an international perspective when discussing innovation technological adoption and organizational adaptation?[1] Four factors may explain why. The first factor is the internationalization of trade and business. For instance, several countries have entered or are currently preparing to enter into free-trade agreements (e.g., Australia-New Zealand, United States-Canada, United States-Mexico, United States-Israel) and in 1992 the Common Market will eliminate any trade restrictions between its member countries. Trading blocks thus formed might impose barriers limiting market access for firms not located in participating countries for commodities such as agricultural products, steel and computer chips. To illustrate, a Swiss firm may need to open a subsidiary in a Common Market member country, duefully incorporated under that country's laws. Such a move will allow the firm to take full advantage of the liberalization of the Common Market (e.g., trade and participate in government sponsored programs) while otherwise its market access may be limited.

The second factor evolves out of the first: internationalization of trade means managing cultural diversity as effectively as possible. For instance, the Swiss firm's subsidiary in West Germany may recruit and employ any engineer or scientist who is a citizen of one of the European Community member countries without any employment restriction (no immigration or work permit necessary), since these individuals are free to move wherever their work place may be located within the European Community. Thus, the firm's workforce is probably somewhat multicultural. In such a work environment of cultural diversity, effective innovation hinges on the people having the necessary habits, skills and styles from which they can construct the appropriate "strategies of action" to manage adaptation (Swidler, 1986). Another example leading to cultural diversity in an organization's workforce is a joint venture between, for instance, a Japanese and a U.S. firm. The joint venture has both Japanese and U.S. employees who must have the tolerance to understand and accept each other's different cultural values and habits in order to work successfully within these parameters.

The third factor stems from the previous two in that a firm's workforce will have acquired different knowledge and skills through formal and informal education. Considering the international workforce or even the cultural variety within a country (e.g., ethnic groups and minorities) the firm must ensure that its employees either already possess or learn a similar set of skills, knowledge and strategies of action through training.[2] The training will assure successful technological adoption and organizational adaptation by enabling employees to work together and understand each others actions, styles and habits. Moreover, it will give employees the new skills needed to make effective use of innovative technology (e.g., Gattiker, 1990).

The last factor is that labor laws, codes and other government imposed rules and regulations will affect a firm's decision-making process about possible technology adoption and organizational adaptation. For instance, the labor law in one country may require that the firm and workers form a committee that discusses and determines how redundancy effects should be dealt with before new technology will be introduced.

Technological adoption necessitates subsequent organizational adaptation to make effective use of new technology. This paper will concentrate on developing a theory[3] offering research strategies to further investigate the interrelationship between culture, training and effective technological innovation as well as organizational adaptation. The theory draws on several disciplines: psychology, sociology and management. It is hoped that this theory will increase our knowledge about how to train individuals to manage innovation and technological adoption as effectively as possible.

DISTINGUISHING THE DIMENSIONS AND ELEMENTS OF THE TECHNOLOGY ADOPTION AND ORGANIZATIONAL ADAPTATION THEORY

If a comprehensive theory is needed, what should the dimensions be? It seems that two dimensions must be added or given particular attention by employees and firms alike: the employee's level of cognitive ability[4] and the macro and micro aspects of cultural stability.[5]

What kind of theory can be developed within the above two dimensions? Further conceptualization of what is included in the two dimensions is necessary before this question can be answered thoroughly. One way to facilitate this conceptualization is to search for general variables which are primarily *culture-free* and *timeless* continua (Hage, 1972, p. 10). Such general variables are universal and easy to work with; more important, they help us recognize that many variables are needed to explain social phenomena. The approach used in developing general variables in this paper has been to identify the dimensions—in this case, level of cognitive ability and cultural stability—and determine the general variables.

Each general variable listed in Table 1 has several elements to it which need to be researched further. *Element* is a primitive term used to define classes of a phenomenon (Hage, 1972, pp. 28, 120). Each general variable and its respective elements will be discussed in more detail below. Why would one choose both cultural stability and cognitive ability to develop a technology adoption and organizational adaptation theory? First, research in high technology firms has shown that cultural stability or instability is an important component for a firm's success in fostering innovation. For instance, Perry and Sandholtz (1988) reported that cultural instability in a high technology firm tends to encourage innovation and organizational adaptation while cultural orthodoxy reduces the tolerance for new ideas and, therefore, hinders innovation. For example, in an orthodox and very stable culture, such as Russia before *perestroika,* cultural diversity was suppressed and adoption of new technologies, innovation and also the necessary organizational adaptation was not especially appreciated, supported nor rewarded.

Conversely, if individuals are used to and accept cultural instability, they may also be more tolerant toward skill development, as well as innovation and adaptation by the firm. This tolerance toward learning new skills will greatly facilitate the formal or informal training needed for individuals to acquire the tools, habits and new skills needed to succeed. If the training is in a formal setting, however, effectiveness of such efforts depends upon students' ability to learn. High cognitive ability in the workforce may not by itself ensure successful adaptation to new technology, but together with

Table 1. Technology Training: Dimensions and Their General Variables

Level of Cultural Stability	*Level of Cognitive Ability*
Technological Adoption and Organizational Adaptation	Training
Structure of Work and Work Processes	Skills

a willingness to be open to change and to accept that action strategies may have to be altered to make better use of new opportunities, it is vitally important.

What distinguishes this model from others? Most conceptual approaches for explaining culture's impact on organizations have been put into a national vacuum. Thus, how leadership, human resource selection and reward systems affect organizational culture and vice versa is described within one culture or nation (e.g., the U.S.) without considering that the internationalization of business limits such an approach. Moreover, training and management development is usually described within one culture, again ignoring the challenges and opportunities for such an approach in an international context.

Another important distinguishing factor of the model outlined in Figure 1 is that both cultural stability and cognitive ability are dimensions whose variables and elements interrelate and affect each other. Thus, as within an organic system, changing the level of one variable will eventually affect the other parts of the system. In most models, technological adoption and subsequent organizational adaptation is not seen as being part of an open and organic system (Scott, 1981), but instead as a static and often closed system.

The last distinguishing factor for the approach outlined here is that most recent models have concentrated on explaining the necessary processes for creating an environment fostering innovation, with very limited focus upon human resource management's interrelationship with potential success in this area. Thus the model outlined in Figure 1 represents an open-system approach to technological adoption and organizational adaptation taking the internationalization of business (cultural stability) and also human resources (cognitive ability) into consideration, thereby distinguishing itself from other more limited attempts.

Figure 1 outlines the theory and depicts it as an open system with two dimensions, namely cultural stability and cognitive ability. Moreover, the figure illustrates the various elements for each dimension of the model and their subsequent variables. In the sections below both dimensions of the theory, namely the level of cultural stability and the level of cognitive ability, and also their general variables and elements as well as interrelationships will be discussed in more detail.

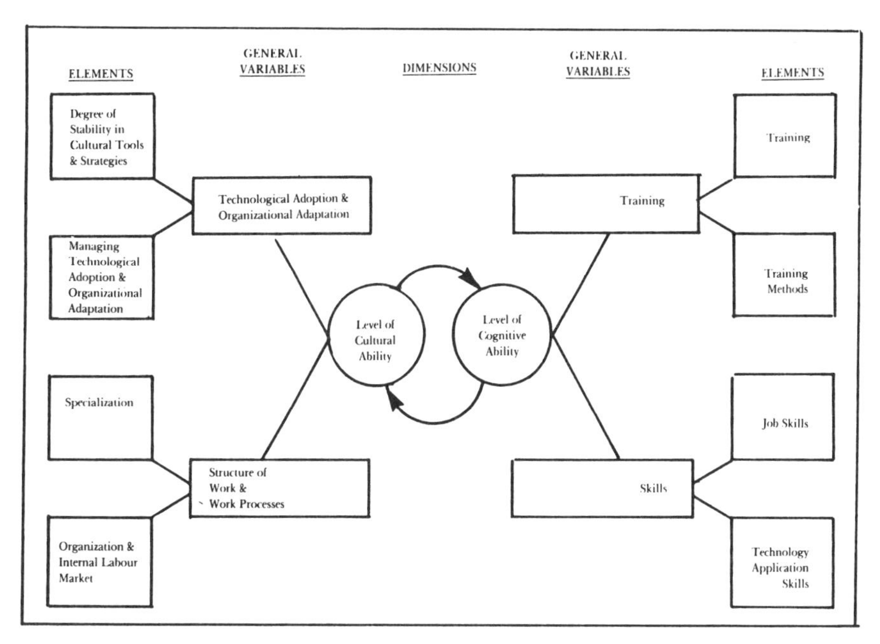

Figure 1. The Components of a Basic Technological Adoption and Organizational Adaptation System

Cultural Stability: General Variables and their Elements

Both a country and an organization's culture encompass symbolic vehicles such as meanings, beliefs, ritual practices, work forms and ceremonies. These also include informal cultural practices such as "the grapevine," gossip, stories and the organizational rituals of daily work life. According to Swidler (1986), these symbolic forms are the means through which action strategies are shaped. Culturally-shaped skills, habits and styles explain what is distinctive about the behavior of organizations, groups and societies. Cultural stability is defined as an environment that sustains existing strategies of action, while instability is inherent in change that causes new patterns of action to evolve.

How does cultural stability relate to managing the firm? As mentioned earlier, based on their study of a high technology firm, Perry and Sandholtz (1988) concluded that, interestingly enough, a "liberating" organizational form encourages innovation in a firm. The culture is therefore not stable, but instead in constant flux and tolerant of internal diversity (strategies, habits and values), thereby encouraging innovation. The firm believes in innovation and, most important, social pressure rewards innovation, while the lack thereof may result in dismissal. Diversity may manifest itself by such simple means as having people come to work in various types of clothing (suit, shorts, baseball cap and sneakers) based on their preference (Goldstone, 1987). Table 2 lists the two general variables and their respective elements which are contained in the cultural dimension of the theory presented earlier (Table 1).

Table 2. Cultural Stability: General Variables and Their Elements

General Variable	*Elements*
Technological Adoption and Organizational Adaptation	Degree of stability in cultural tools and strategies (*V1*) Managing technological adoption and organizational adaptation (*V2*)
Structure of Work and Work Processes	Specialization (*V3*) Organization and internal labor market (*V4*)

Technological Adoption and Organizational Adaptation

Organizational adoption as used in this paper refers to the firm's decision to adopt a new technology or innovation. Adoption may occur via a continuous or discontinuous process. The latter usually represents a radical change, while continuous adoption is a gradual process. An example of discontinuous adoption might be a situation wherein an office employee, who has never worked with a computer, arrives at work one day to find a personal computer connected to a local area network at his or her desk. Hence, tasks done manually yesterday should be done with the computer today. A continuous adoption is illustrated by a case where an employee changes from a mainframe terminal to a networked personal computer. Hence, technological adoption requires organizational adaptation to make effective use of the technology. Organizational adaptation could be as simple as preparing a work place (ergonomic work table for using a computer) to rearranging organizational structure.

The term *organizational adaptation* is used in a number of ways in current literature, ranging from referring to reactive or proactive change (Miles & Snow, 1978) to more specifically denoting reaction to environmental changes (Astley & Van de Ven, 1983). Moreover, organizational adaptation is also used to describe the alignment of organizational capabilities with internal contingencies, such as the employees' attitudes toward technology. In this context, the term refers to both proactive and reactive behavior on the part of the firm with the intention to align its organizational capabilities with internal and external contingencies.

The distinction between adoption and adaptation lies in the fact that adoption usually represents the installation of new technology (e.g., in the office or factory). In contrast, adaptation is the organizational system's attempt to adjust to the technology. This adjustment may result in changes in work structure and processes. Job descriptions may be altered and new positions may be formed. The flow of information and the decision-making process may change due to the adoption of new technology. Thus adoption is the first step in an often timely process by the organization to adapt the technology and its own structures and processes in such a way that they fit and, therefore, allow the effective use of new technology.

Organizational adaptation is often caused by product or process innovation that was triggered internally (e.g., R&D unit), externally or by a combination of both. In either case, innovation usually requires adaptation by the firm to respond or make effective use of it. The general variable, technological adoption and organizational adaptation, is obviously important to managing change for the future survival of the firm.

Cultural Tools and Strategies

As discussed in the above section, individuals must acquire certain tools and action strategies in order to accomplish desired outcomes. Based on these varying skills, technological adoption and organizational adaptation is managed differently. One of the areas that has fascinated North American researchers in the last decade is Japanese management techniques and philosophy (Tanaka, 1988). Unfortunately, some research indicates that these may not be applicable in other countries because the necessary action strategies could not develop out of a North American or European worker's "tool kit" (Hofstede, 1984; Swidler, 1986; Tanaka, 1988).

To manage cultural diversity in one's workforce and promote change in cultural values, strategies and habits effectively, potential sources of diversity need to be identified and understood. It seems that four sources for cultural diversity can be identified. First, cross-national differences (political climate, religion and other factors) between, for instance, Japan and the United States must be recognized and dealt with before establishing a joint venture (Osborn, Strickstein, & Olson, 1988). Second, geographical/regional culture (e.g., language, cuisine, life style) may affect the firm's success. For example, in Geneva, not only the language (French) is different than in Zurich (Swiss German), but lifestyles and cultural values also differ significantly (Hofstede, 1984). Third, intracultural differences between ethnic groups (Chinese vs. Polish immigrants in the U.S.) and between the majority and minorities, such as Maoris in New Zealand and Hinus in India, will further increase the cultural diversity in a firm. Fourth, the firm must deal with the professional or trade culture. A joint venture between an American firm (in France) and a Japanese firm (in West Germany), may have to confront the fact that the French (U.S.-trained) managers are trained in public administration (finance and accounting) (Fligstein, 1987), while the German (Japanese trained) managers have an engineering (production) background. Last but not least, the two firms' organizational cultures (e.g., when attempting to make a takeover or start a joint venture) may be totally different. For example, when Asea (Sweden) and BBC (Brown-Boveri Company, Switzerland) merged in 1987, two different cultures led to a great diversity in action strategies, values and habits. Thus the CEO of the new Asea/BBC conglomerate had to hold numerous talks with unions, employees and other groups to explain what the new organization would be all about (e.g., future strategy, areas of growth, job security, risk taking and management style).

To manage the above four sources of cultural diversity effectively, Nissan decided to bring all its first-line supervisors in its first North American plant to Japan for several weeks. This approach enabled the Americans to learn about Japanese management, quality circles and many other things that supposedly were important habits, skills and action strategies accounting for part of Nissan's success. Moreover, the Americans learned about diversity caused by their own regional (South vs. Midwest) and/or intracultural (ethnic group) differences and Nissan's organizational culture. This approach does not eliminate cultural diversity, but it increases tolerance for it by offering the individual the opportunity to *observe, experience, understand* and *learn* to work with different cultural habits and strategies.

Managing Technological Adoption and Organizational Adaptation

In a prescient early observation, Leavitt (1964) suggested that important changes can be brought about in an organization through alterating its technology. Nonetheless, change risks disturbing the psychological contract or status quo for employees who have learned an earlier set of rules (Larwood, 1984, p. 213). Since people generally wish to protect the systems they have found to be successful and want to avoid potentially risky restructuring (Klein, 1966), most people cannot be expected to welcome major change.

Technological adoption often requires firms to continuously reorganize and adopt their organizational processes. For instance, new hardware and software developments lead to changes in the use of computer-based information technology in organizational settings (Gattiker, 1984). Technological developments have led to the transformation of the manufacturing sectors in many countries. Flexible manufacturing systems (FMS) have changed the production process as well as work structure. Such innovation inevitably affects a firm's workforce. Employee acceptance of such changes must be secured in order to facilitate effective use of the technology (Gattiker, 1987c).

Organizational adoption may be facilitated by the participation of the workforce in the decision-making process. For instance, West German laws outline exactly how workers and their union representatives must be informed about technology and organizational changes due to adoption (e.g., Osterloh, 1986). Despite this, an extensive case study by Wilpert (1986), which looked at five West German companies, concluded that, unfortunately, the legal stipulations were inadequate to set a comprehensive framework for the decision process. Wilpert also suggests that agreements between unions and firms to consult and bargain throughout the process of introducing new technology would ensure that employees are informed and involved in the process. This could lead to greater acceptance of technological adoption and smoother organizational adaptation.

Participation can be accomplished through legal requirements or informal, voluntary participation. For instance, while West German and French firms are required by law to let their workforce participate in the decision process when it comes to technology and the necessary organizational adaptation, American and Japanese firms use the voluntary model. Still, strategy differences within each group of countries, such as those between the voluntary models followed by the United States and Japan, are common. Worker participation in the decision-making process about technology adoption at Ford may, therefore, largely be based on union contracts, while at Toyota technology adoption is being discussed thoroughly within quality circles and only initiated after some kind of consensus has been achieved. Furthermore, how participation may be implemented and used is most likely going to differ between two firms in one country (Ford vs. Chrysler). Each approach toward participation (legal or voluntary), however, does achieve involvement of the workforce in the decision-making process and thus is valuable.

In an organizational culture that tolerates diversity, it is possible that even where legal participation and codetermination is stipulated, informal channels may be used first. Such an approach helps management to encourage the use of various solutions to problems, thereby increasing the creativity of the workforce. Before a solution becomes the norm, however, it has been developed based on informal participation to fit the employees' and the firm's needs, thereby making the participation steps required by law a rubber stamp process (Gattiker, 1988b).

Structure of Work and Work Processes

In recent years, much attention has been paid to the phenomenon of the *deskilling* of craft workers (Braverman, 1974; Hall, 1986). Deskilling is seen as an act of capitalism that attempts to transfer control of work to management by depriving the employee of his or her skill. Integrated, self-controlled craft work is replaced by centralized design, standardized procedures and products, and the fragmentation of skills into detail work in unskilled, specialized roles. Nonetheless, as will be shown below, Braverman's thesis that the workforce is becoming increasingly unskilled and homogenized is not necessarily conceptually and empirically founded (Francis, 1986, chap. 5).

Elements of Specialization

The elimination of certain crafts has led to the creation of different occupations. Technological developments account for some of these changes. Nevertheless, these developments may have led to such undesirable effects as the extreme specialization of certain occupations. Some occupations have a narrowly defined set of responsibilities and any jobholder must have certain skills in order to work with the new technology. For instance, a study by Shaiken, Herzenberg and Kuhn (1986) found that even in small-batch production, there was little sign of the new "craft worker" described by Piore and Sabel (1984). Moreover, a premium placed on quality and fast delivery would require that technology and shop-floor skills be used in a complementary fashion, which these authors did not find. Instead, programming changes were not made on the shop-floor but by a separate programming department. A similar fragmentation was discovered by Kraft and Dubnoff (1986), who looked at computer software workers at the leading edge of the computer revolution. These results are not universal. In a contrasting study, Attewell (1987a) used quantitative industry-wide data on insurance occupations and found no aggregate deskilling. Instead, he even reported in one study that skill levels increased due to office automation for clerical type positions in insurance offices (Attewell, 1987b). Similarly, in West Germany and Sweden, up-skilling has resulted from technology-induced adaptation in firms and industry (e.g., Brumlop & Juergens, 1986; Van Houton, 1987), thus suggesting that technology may not necessarily lead to deskilling as suggested by Braverman (1974). These data suggest that technology's effects upon skills might depend on the type of work industry and firm or country involved to mention a few variables.

Although the above suggests that cross-national differences may account for the various outcomes reported in studies, three interrelated factors may better explain why in some cases up-skilling and in others deskilling may occur due to technological change. First, labor market structure differences may be apparent. For instance, internal labor markets (see the next section) in Sweden and West Germany often guarantee employees that quality of work life should improve with new technology (Betriebsrat, Max Planck-Institut fuer Bildungsforschung, 1988). Thus deskilling is not a viable option (Brumlop & Juergens, 1986; Van Houton, 1987). Second, up-skilling of jobs with the help of technology could be one strategy for fighting high labor costs (UBS, 1988), thereby increasing productivity of capital. In contrast, low labor costs, as for certain positions in the United States (e.g., hired under contract at minimum wage without any fringe benefits and no opportunity to acquire seniority rights), may encourage further deskilling to reduce wages and facilitate the hiring of cheap labor (e.g., illegal immigrants). Third, formal education received by employees and the skills needed may determine how much up-skilling could be a viable option. If people with certain skills are not available and workforce tenure is relatively short, it may be more economical for a firm not to invest in training but, instead, keep the skills requirement for performing a job as low as possible. Furthermore, illiteracy and lack of understanding of basic mathematics may prevent an organization from up-skilling certain jobs unless its employees acquire the skills in a remedial program.

The above shows that further deskilling or up-skilling with the help of technology depends upon the three factors described above and suggests that Braverman's thesis is neither right or wrong but too general to explain the process adequately.

The Organization and its Internal Labor Market (ILM)

For the purposes of this discussion, the central idea of the internal labor market approach is that both managers and workers in the large and medium-sized firms, as they try to maximize opportunities for profits and wages, attempt to reduce uncertainties arising from the market environment. For managers, a major concern is the uncertainty over whether employees will stay with the firm. For instance, virtually all computer programmers in North America used to learn their trade either in school or in private training institutions, while Europeans learned through apprenticeships with a mixture of organizational training and outside schooling. At that time, programmers' loyalty was to their profession but not necessarily to their firm. Turnover was relatively high since these individuals had no difficulty finding other jobs and transferring their skills. As well, their wage demands had the potential to interrupt internal wage scales. Therefore, some companies started offering a more truncated training program to qualified applicants within the firm. This, in turn, stabilized the internal pool of programmers since they could no longer transfer their skills as easily as they had before (Osterman, 1986). This trend is continuing and increasing in other positions, not only as a result of technological innovation, but primarily because of companies' attempts to stabilize their internal labor markets.

For employees, the major point of uncertainty is that unfavorable labor market conditions, such as high unemployment, will lower the costs of their dismissal and replacement by new workers. Thus workers also have an interest in institutionalizing the administrative rules that govern internal job ladders and wage and salary increments linked with promotion. Employees do attempt (especially through trade unions) to enforce principles such as seniority as the primary criterion for determining the order of promotions and temporary lay-offs (Pfeffer & Cohen, 1984). To the extent that employees are successful in enforcing such rules, uncertainty of employment is differentially distributed across the workforce and the actual or potential beneficiaries become more likely to cooperate in the production process (Stark, 1986). The efforts made to reduce uncertainties by both firms and employees, therefore, take the form of bureaucratic rules which are ILMs' distinctive trait (Osterman, 1984, pp. 2-6).

This means that ILM structure can affect technological innovation and adaptation efforts. For example: work roles may prevent employees from doing certain jobs, and implicit employment guarantees may prevent lay-offs. This relates to culture because organizations use different approaches to manage ILM based on their culture and nationality. Piore (1986) concluded that although American firms have greater employment flexibility under the law, cultural parameters such as union contracts and implicit employment guarantees may force them to lay-off workers according to seniority as a last resort. In Western Europe, laws outline procedures that limit a firm's choices. Nevertheless, the final outcome for labor market flexibility is similar, despite the different cultural strategies used.

A firm's internal labor market also limits its choices of how to manage technological adoption and organizational adaptation. This means that bureaucratic rules can put an organization into a straightjacket (Betriebsrat, Max-Planck-Institut fuer Bildungsforschung, 1988). With new technology, this could mean that individuals cannot be shifted easily from one job to another within a firm, nor can job duties and skill requirements be changed without major adjustments to job descriptions and union bargaining contracts.

Summary and Conclusion

We have suggested that there are specific relationships between the general variables identified above. Table 3 represents a way of schematizing these relationships. Hage (1972, chap. 4) argues that linkages between elements of a theory must be specified with theoretical statements. These are phrases that indicate why, whereas operational linkages explain how. Hage states that operational linkages make a theory measurable and should explain if the linkage is linear, a curve, or exponential,[6] thereby simplifying data-processing and analysis.

Table 3. Theoretical and Operational Linkages Between Elements of Culture

Technological Adoption and Organizational Adaptation	*Structure of Work and Work Processes*			
	Specialization (V3) Theoretical Linkage	*Operational Linkage*	*ILM (V4) Theoretical Linkage*	*Operational Linkage*
Degree of Stability in Cultural Tools and Strategies (*V1*)	A higher degree of cultural stability increases the level of specialization of the workforce	Positive linear correlation with limit and different coefficients	A higher degree of cultural stability increases the level of bureaucratic rules for a firm's ILM	Positive linear correlation with limit and different coefficients
Managing Adoption and Organizational Adaptation (*V2*) increases the level	Participation based on legal constraints of specialization of the workforce	Power curve with limits type of voluntary	Participation based on labor laws or some with limit and agreement increases the level of bureaucratic rules for a firm's ILM	Positive linear correlation different coefficients

Table 3 illustrates the different theoretical linkages between the elements of the two dimensions. The operational statements specify the coefficients that represent the relationship between the two variables. For instance, *V1* and *V3* have a positive linear correlation with different coefficients and a lower/upper limit. How can this be interpreted? Every country and its culture experience some degree of stability that is represented by the lower limit. Additionally, stability tends to lead to work forms or jobs that become more specialized (e.g., extensive technological change limits job specialization and skills stagnation) as the culture becomes more stable and immune to change. The different coefficients mean that the relationships can be made up of different ratios between *V1* and *V3*, ranging from .01 to a perfect 1.00 relationship.

Power is a coefficient that is constantly changing (e.g., power curve). Adoption and adaptation (*V2*) and specialization (*V3*) have a relationship that can change and is not linear (Hage, 1972, pp. 100-106). The lower limit suggests that some innovation and specialization is inevitable for all organizations. Furthermore, technological adoption and organizational adaptation can be continuous, but their frequency and the level of specialization that can be reached is limited. Exactly where the upper and lower limits are, as well as what power coefficients therefore apply, still needs to be tested.

The relationship between *V2* and *V4* is positive and linear but different coefficients are possible. Hence, the more technology adoption and organizational adaptation a firm experiences, the more likely its workforce will push for some sort of participation in the decision-making process. Increased participation does, however, increase bureaucratic rules since numerous factors must be considered before the firm may introduce a new technology. There may exist a maximum as well as a minimum, but most firms will be somewhere in the middle. The correlation may also differ based on such factors as cooperation and trust between management and employees. Distrust may increase the tendency for rules and regulations.

Table 3 lists the basic operational linkages only. As a result, it provides only a framework for future research. The table also indicates that bureaucratic rules for ILM increase with increased levels of cultural stability (*V1* and *V4*). Cultural stability tends to increase the number of laws affecting organizations as well as employees. In both cases, the operational linkage suggests a positive linear relationship. Again, the different coefficients, as well as the upper and lower limits, need to be identified by future research.

Level of Cognitive Ability: General Variables and Their Elements

There is very little research available that deals with the effects of cognitive ability on technology training. Educational psychologists have argued that a person's ability will influence how he/she will learn and what performance outcomes should be expected (Lepper, 1985; Snow, 1986). For example, Gattiker (1987a) found that students with less cognitive ability (e.g., lower grade-point average) did better in a course teaching micro-computer skills if they had previously taken a course in computer science. Another study found that individuals with less cognitive ability benefit most from additional time spent on learning exercises using new technology (Gattiker, 1987b). One reason could be that additional hands-on experience, acquired when practicing one's skills on the computer (or having previously attended another computer course), speeds up one's automated processes (e.g., using a software package without having to use the manual) and thereby positively affects one's overall performance. Nonetheless, limited cognitive abilities may prevent total elimination of the performance gap between ability groups.

Organizations will have to train individuals who have various levels of ability. Some employees may be functionally illiterate, whereas others, such as R&D personnel, may have graduate education in engineering or other areas. In either case, it is obvious that training methods must differ. An additional concern must be the individual's past job experience. For instance, Kohn, Schooler, Miller, Miller, Schoenbach and Schoenberg (1983) reported in their longitudinal study that an employee's intellectual flexibility is influenced by his or her experienced job complexity. High job complexity will increase intellectual flexibility while low complexity will decrease intellectual flexibility. The degree of complexity is synonymous with the variety of skills the job holder should possess in order to complete the duties and responsibilities involved. The research by Kohn and colleagues suggests that past job experiences as well as formal education will influence a person's current level of ability. This will affect one's training ability as well as skill level (depth, range and application). Table 4 lists the two general variables and their respective elements that are contained in the ability dimension of the theory presented earlier (Table 1).

Table 4. Level of Cognitive Ability: General Variables and Their Elements

General Variable	*Elements*
Training	Training content (*V6*) Training method(*V7*)
Skills	Job skills (*V8*) Technology application skills (*V9*)

Training

Training Content

Knowledge theorists argue that an important prerequisite for problem solving is the ability to access knowledge when needed. Acquiring knowledge relevant to one's job situation, therefore, is no guarantee that individuals will activate that knowledge in a relevant situation (Dooling & Lachman, 1971). Educational research indicates that activation is easier for people if their previous experiences provide a basis for producing relevant schemata (e.g., Gick & Holyoak, 1983), and if learning activities help individuals to experience problems and recognize the usefulness of their own knowledge for solving them (Adams, Perfetto, Yearwood, Kasserman, Bransford, & Franks, 1985).

Training must be designed in such a way that individuals can access newly acquired knowledge in relevant job situations. An individual should be capable of using the technology innovatively in his or her job. This suggests that employees should be able to access cultural skills and strategies applicable to situations such as international contract negotiations. Also, R&D engineers must be able to work together with production workers whose cultural skills are very different. In this context, learning another language in a classroom should be complemented with an opportunity for the individual to live in that country, thereby learning the cultural nuances of words as well as their various meanings.

Training Method

Training should focus on allowing the individual to master challenges during the period of learning. The overall goal is to increase the employee's competence in a particular area, while a learning goal focuses on activating his or her own ability to overcome obstacles, thereby increasing learning ability (Dweck, 1986).

New technology has opened new avenues for training. For instance, computer-aided learning (CAL) has become popular. CAL has three distinct dimensions:

1. Drill-and-practice applications, in which the instructor presents material to students by conventional means and the students practice the new skills using the computer during class or on their own time.
2. Tool mode, in which the student uses the computer to perform certain tasks utilizing software and statistical packages with and without time constraints.
3. Tutee, in which students give the computer directions in a programming language it understands, such as Pascal.

Of course, in addition to CAL, there are more traditional approaches to learning, such as studying a manual or using a laboratory setting for learning exercises. Furthermore, video disks are also used to train workers. The issue here is to determine not which of the methods is better, but what different methods of training should be used to increase learning for employees with different levels of cognitive ability. Research by Gattiker and his colleagues found that university students with low cognitive ability gained most from CAL when trying to acquire computer application skills (e.g., Gattiker, 1987a; Gattiker & Paulson, 1987). Depending upon an individual's learning style, however, teaching styles must differ considerably. No matter what kind of training method is used, the ultimate objective is to increase the individual's competency.

A teaching method that presents information the way the individual prefers to process it is an important factor to successful training. Gregorc (1982) developed a taxonomy of learning that consists of four cognitive styles: 1) logical sequential, 2) abstract sequential, 3) abstract random, and 4) concrete random. Each label represents a style of ordering and processing information that dominates an individual's approach to learning new skills. Effective learning requires not only a match between the individual's learning style and the teaching style but also, most interestingly, a match between the technology and the styles. For example, logical sequential learning is most successful when CAL follows a sequence of logical steps to guide the individual to a higher level of mastery. For the concrete random individual, however, such techniques might fail because the individual likes to study the problem from different angles, not necessarily following a logical sequence of steps. The above exemplifies that a person's learning style and preference in processing information must be considered when trying to design a training program for a firm's workforce.

Job Skills Facilitating Effective Use of New Technology

The discussion of skills also involves the range and depth issue. Specialization leads to a smaller range but greater depth of skills. For instance, a nurse specializing in gynecology will have a great depth of skills in this area but may have a limited range of skills in emergency care. This suggests that there is a linkage between skill requirements for certain positions and specialization of employees. Positions that require highly specialized skills of limited range and great depth may be difficult to transfer into unless training is provided. Similarly, a transfer out of such a position may also require additional training to smooth the transition.

Job Skills

Sociologists have claimed that jobs have inherent skill requirements. Thus, the primary mechanism through which skilled people select their work appears to be long-term occupational shaping rather than immediate or substantial changes in the structure of their work (e.g., Kohn et al., 1983; Spenner, 1983). Professional and craft workers offer different skills and choose different jobs that make use of those skills (Kohn et al., 1983). Furthermore, the comparison between professional and craft jobs indicates that there are major differences in the content and length of training. The former emphasizes and uses theory while crafts stress and use hands-on experience (Hall, 1986, p. 68).

In the past, we have usually distinguished between industrial (blue-collar), salaried (white-collar), craft and secondary employment sub-systems (Osterman, 1986). At this time, however, what interests us most is whether the future will offer employment subsystems including the following work types: (1) knowledge worker, (2) skilled, (3) semiskilled, and (4) low-skilled employee. The knowledge-worker category would include managers, R&D scientists and engineers—professionals who acquire, process and interpret information from a variety of sources to solve problems. Professionals who diagnose problems, such as doctors and lawyers, are also included in this category. The usual distinction between blue and white-collar workers has been dropped. Instead, this system distinguishes between skilled workers—including potters, nurses and teachers—and semi-skilled workers—computer operators, bank tellers and so forth. Skilled workers acquire skills that are not firm-specific. These employees thus have more market power than their semiskilled counterparts who have firm-specific skills. Low-skilled employees include mail room staff, messengers, fast-food clerks, and gas station attendants. This fourfold classification system is far from earlier efforts to understand subcategories of work; it differs in that it tries to capture some of the technological adoption and organizational adaptation developments that have led to different positions and jobs.

The benefits of this classification system can be shown by contrasting it with a traditional system of classification. Computer software specialists design, write and modify the instructions that make computers work. Using a traditional classification system, one would say that this is both a white-collar (Osterman, 1986) and professional job (Hall, 1986, p. 68). This categorization, however, is inadequate to distinguish between different types of software positions. Job content and the degree of control exercised determine which of the first three categories (knowledge, skilled, or semiskilled work) the job will be put into. Kraft and Dubnoff (1986) found that, for instance, software specialists doing maintenance/application work and system programming were semiskilled because their skills were firm-specific, job content was narrow and control limited. Skilled software positions, however, require designing applications based on specific customer needs and demand frequent interaction with customers and peers. Software employees in the knowledge category had to plan, manage and make decisions about software and hardware purchases and sales that affect the lower level software positions. This example illustrates the usefulness of the four-category system presented in this paper.

How will these four job categories apply with the introduction of new technology into the work world? Rapid change and its inherent instability forces people to adjust. For instance, technological changes require learning new skills and change requires intellectual flexibility and a person's capability to face complexity (Kohn et al., 1983). Training and retraining to acquire new skills may become the single most important factor in adaptation by helping people to move into new positions.

Technology Application Skills

These skills vary according to the technology used by the individual, which may range from computers to lasers. For office workers, application skills for work stations usually include a person's ability to use word processors, spreadsheets, database managers, some statistical packages, and programming and business-related software such as graphics (Jones & Lavelli, 1986). Gattiker (1987c) has argued that a person with the above skills has a first level of literacy in a technology. A systems programmer developing an expert system for oil exploration or a fighter pilot, however, requires skills at the professional level.

Technological innovation has created different types of technology application skills. For instance, research about computer-numerical control

(CNC) machines and flexible manufacturing systems (FMS) indicates that planning responsibility and autonomy are often absent from the shop-floor (Shaiken, Herzenberg, & Kuhn, 1986). Managers and skilled workers in the office who have some shop-floor experience contribute the knowledge needed to design and plan new products and implement their production. This results in deskilling of the formerly skilled crafts worker on the shop-floor. He or she is left to operate a machine, which is semi-skilled if not unskilled work. Shaiken, Herzenberg and Kuhn (1986) stated that a more broadly skilled workforce emerging from FMS and CNC production holds more promise in theory than in practice (Piore & Sabel, 1984). The machine operator can often override the computer program during the manufacturing process; however, actual program changes are done by someone else. This would indicate that technology application skills for semiskilled or low-skilled employees are narrow in focus and scope and further encourage specialization.

Bikson and Gutek (1983) found that office workers' application skills for work stations were not adequate for them to make full use of the computer-based technology in their jobs. Instead, their training was so job specific that it prevented them from developing applications on their own using standard software. Similar results were reported by Verdin (1988). She found that some managers were not satisfied with personnel information systems because they did not feel adequately trained to use them effectively. It appears that employees often do not have adequate technology application skills for them to make effective use of technology. Furthermore, their skills may be limited in scope and decision making may not be required, so these skillful employees become deskilled operators. Nonetheless, even though the above developments (e.g., deskilling) may occur, cross-national differences as outlined earlier in this paper (see Elements of Specialization) may in part account for the sometimes negative skills outcomes attributed to technological adoption.

Summary and Conclusion

The sections discussed above indicate that there are specific relationships between the elements of the general variables. Table 5 presents a way of schematizing these relationships. Again, theoretical linkages as well as operational ones are identified.

The relationship between the learning goal (*V6*) and a person's job skills (*V7*) is curvilinear. This means that up to a certain point, one's competency level increases along with the number of skills. After the maximum point has been reached, the number of skills may still increase but one's competency in using them may deteriorate. One reason for this could be a person's limited cognitive ability to store all of this different information.

Table 5. Theoretical and Operational Linkages Between Elements of Cognitive Ability

	Skills		*Work Process*	
Training	*Job Skills (V7) Theoretical Linkage*	*Operational Linkage*	*Technology Application Skills (V8) Theoretical Linkage*	*Operational Linkage*
Training Content (*V5*)	Easier access of knowledge increases the range and depth of job skills	Positive linear correlation with limit and different coefficients	Easier access of knowledge increases the range and depth of technology application skills	Positive linear correlation with limit and different coefficients
Training Method (*V6*)	Increased competency improves range and depth of job skills of the workforce	Curvilinear linkage with different coefficients	Increased competency improves the range and depth of technology application skills of the workforce	Curvilenear linkage with different coefficients

The different coefficients simply mean that the curve and its peak can change for each employee or work group. The theoretical linkage between variables 5 and 7 suggests that easier access to knowledge increases the level and depth of skills. Accessing the knowledge easily and regularly in one's job is the most effective way to retain skill and proficiency (e.g., Kohn et al., 1983). Training with the objective of improving an individual's competency in using the technology will enable the individual to increase the range and depth of his or her skills (*V6* and *V8*) and use the technology more effectively on the job. Again, the relationship is curvilinear because the individual can increase his or her competency for a range of skills only to a certain level, and beyond it, competency will decrease. Increasing one's range of skills may reduce the in depth knowledge about applying various skills to one's job.

THEORETICAL AND OPERATIONAL LINKAGES BETWEEN CULTURAL STABILITY AND COGNITIVE ABILITY

The theoretical perspective just described should be recognized as merely a skeletal framework. The theoretical linkages as well as the operational definitions needed to test the theory have been identified. The research necessary to describe the pieces themselves still needs to be done, and the dynamics of how the pieces of the theory interrelate to one another awaits testing. I hope that the dynamics will be examined by later researchers.

Table 6 outlines how training content and training method relate to the degree of stability in cultural tools and strategies as well as managing adoption and organizational adaptation. For instance, increased competency (*V6*) improves the ease with which technological adoption and organizational adaptation (*V2*) triggered transition can be managed. Thus if individuals know how to make effective use of new technology due to their training, they will be more likely to adjust to the change and feel less threatened by the outcome.

Table 6. Theoretical and Operational Linkages Between Elements of Cognitive Ability (Training) and Cultural Stability (Technological Adaptation and Innovation)

Technological Adoption and Organizational Adaptation	*Training*			
	Training Content (V5)		*Training Method (V6)*	
	Theoretical Linkage	*Operational Linkage*	*Theoretical Linkage*	*Operational Linkage*
Degree of Stability in Cultural Tools and Strategies (*V1*)	Less stability in the culture increases the desire for abstractness of training information	Positive linear correlation with limit and different coefficients	The more stability in cultural strategies & tools, the more training must focus on practicing the new skills (hands-on)	Postitive linear correlation with limit and different coefficients
Technological Adoption & Organizational Adaptation (*V2*)	Increased adoption requires more job relevant training to improve access of relevant job skills	Positive linear correlation with limit and different coefficients	Increased competency improves the ease with which technological adoption and innovation triggered transition can be managed	Positive linear correlation with limit and different coefficients

Table 7 proposes the interrelationships between skills and technological adoption and organizational adaptation. It is suggested that limited participation in the technology adoption and organizational adaptation process (*V2*) will reduce the skill levels attained by the firm's employees. One reason for this could be that participation helps employees at an early stage to learn more about new technology and start to understand how it will be used in their work area. Moreover, certain skills are acquired during these early stages (Gattiker, 1988a), which should motivate individuals to acquire further skills early on either through formal (e.g., training) and/or informal means.

Table 7. Theoretical and Operational Linkages Between Cultural Stability (Technological Adaptation and Innovation) and Cognitive Ability (Structure of Work and Work Processes)

Technological Adoption and Organizational Adaptation	*Skills*			
	Job Skills (V7)		*Technology Application*	
	Theoretical Linkage	*Operational Linkage*	*Theoretical Linkage*	*Operational Linkage*
Degree of Stability in Cultural Tools and Strategies (*V1*)	A higher degree of cultural stability decreases the level of cultural tools and skills held by the workforce	Negative linear correlation with limit and different coefficients	A higher degree of cultural stability increases the level of bureaucratic rules for a firm's ILM	Postitive linear correlation with limit and different coefficients
Managing Innovation and Adoption(*V2*)	Limited participation by employees during the adoption process decreases skill levels	Power curve with limits	Participation based on labor laws or some type of voluntary agreement increases the level of bureaucratic rule for a firm's ILM	Positive linear correlation with limit and different coefficients

One of the major points remaining is the linkage between the structure of work and work processes and the skills. Table 8 suggests that a highly structured ILM does increase the structure of training content. Hence, due possibly to skill level changes, training content must be approved to avoid potential job classification changes. Due to the nature of the strict approval process (e.g., union and management), a highly structured ILM will likely increase logical and sequential (step-by-step) training following a preapproved outline. Thus, on-the-job training is less likely. Table 9 outlines the relationship between skills and the structure of work and work processes.

In Table 9 the relationship between *V3* and *V7* suggests that increased specialization improves the depth of a person's job skills in a certain area, but limits his or her skill range. The same applies for technology application skills (*V3* and *V8*). A highly structured ILM tends to result in narrow job descriptions that, in turn, cause the individual to become highly skilled in a narrow area applicable to his or her job. Once again, the same is true in a highly structured ILM for technology application skills.

The above proposes interrelationships between the variables outlined in Tables 2 and 4 and suggests the operational linkages. Most important is the fact that levels, of cultural stability and cognitive ability are assumed to be interrelated. Thus a training or adaptation strategy within a firm must take this interrelationship into careful consideration.

Table 8. Theoretical and Operational Linkages Between Elements of Cognitive Ability

Structure of Work and Work Processes	*Training*			
	Training Content (V5)		*Training Method (V6)*	
	Theoretical Linkage	*Operational Linkage*	*Theoretical Linkage*	*Operational Linkage*
Specialization (*V3*)	Easier access of knowledge increases the range and depth of job skills	Positive linear correlation with limit and different coefficients	The more advanced the ILM, the more structured will training content be	Positive linear correlation with limit and different coefficients
ILM (*V4*)	Increased competency improves range and depth of job skills of the workforce	Curvilinear linkage with different coefficients	The more advanced the firm's ILM, the more logical sequential the training method will become	Positive linear correlation with limit and different coefficients

Table 9. Theoretical and Operational Linkages Between Elements of Cognitive Ability Dimension (Skills) and Cultural Stability (Structure of Work & Work Processes)

Structure of Work and Work Processes	*Training*			
	Job Skills (V7)		*Technology Application Skills (V8)*	
	Theoretical Linkage	*Operational Linkage*	*Theoretical Linkage*	*Operational Linkage*
Specialization (*V3*)	Higher specialization increases depth of job skills but decreases the range of skills	Positive linear correlation with limit and different coefficients	Higher specialization increases the depth of technology application skills but decreases the range of skills	Positive linear correlation with limit and different coefficients
ILM (*V4*)	A highly structured ILM increases the depth of job skills but decreases the range of skills	Positive linear correlation with limit and different coefficients	A highly structured ILM increases the depth of technology application skills but decreases the range of skills	Power curve with limits

CONCLUSION

The model proposed suggests that successful technology adoption and organizational adaptation require a fit between culture (organizational and national) and cognitive ability by the employees. Such a fit will enable the organization to offer the necessary training where it is needed most, thereby allowing employees to acquire the necessary skills to perform well with the new technology. Due to space limitations, the legal environment was not extensively discussed, although it will affect the interrelationship between technology adoption and organizational adaptation profoundly. Culture may be an important factor since labor laws may differ across provinces/ states and most certainly between countries. Thus action strategies employed by a company's subsidiaries must differ due to local labor laws. For instance, in less-developed countries, firms may introduce technology and automatically lay off redundant workers in large numbers. In Canada, if technology adoption results in more than 50 lay-offs within any four week period, the group termination falls under The Canada Labour Code, sections 59.7 - 60.31, and a joint committee consisting of redundant workers and management must be established. The committee will determine severance pay, retraining support and any other compensation to be paid to the redundant workers. If the committee cannot agree, a federally appointed arbitrator will draft an agreement that will be binding for both parties.

Implications for Managers

The most obvious practical implication deriving from this model is the firm's need for an integrated training strategy considering cultural factors. However, popular management literature is of little help here. For instance, recent suggestions in the popular literature point out demographic trends in the U.S. workforce (e.g., in the 1990s the majority of new employees available will be women and black, Hispanic or Asian men) and state that firms must be successful in attracting these employees (Kupfer, 1988). Unfortunately, such information does not mention or discuss the potential effect of these shifts upon cultural stability in organizations. Moreover, how a firm might manage this increase in cultural diversity successfully in combination with training is anybody's guess.

Globalization of business does not only require having subsidiaries in various countries and a global strategy linking them with headquarters. More important is the fact that training strategies need to be developed that differ for various groups of employees (e.g., knowledge workers vs. unskilled laborers) when it comes to technical skills training. It remains to be discovered how the firm can assure that knowledge workers will form an effective team with unskilled laborers. The two groups may not have the same mind-set; that is to say, values, norms and action strategies may be so different that communication becomes difficult, even if they speak the same language.

This paper would then suggest that a firm must offer or support three types of training: (1) job specific training, (2) training to acquire plant/

subsidiary cultural habits and action strategies and (3) training for understanding and using the organizational culture to everybody's advantage.

Depending upon the formal education acquired by the new employee, skill training may differ across countries. For instance, vocational training in Europe may be at a more advanced level than in North America. Moreover, skills and knowledge acquired during formal education may assure a European employee's literacy, thus making such upgrading programs more of a North American phenomenon than anything else. Additionally, skills acquired during vocational training may determine in part what additional on-the-job or off-the-job training may be required to perform well.

The second type of training must be management's attempt to create a corporate culture that penetrates all branches and subsidiaries around the globe. The training may differ across subsidiaries depending upon the cultural differences between headquarters and the subsidiary's geographical location. For example, if the latter is located in China, it may be quite difficult to create an environment that favors risk taking and encourages the development of innovation and rapid adaptation due to differing political and economic environments. Thus, a modified culture fitting the national culture may be a more realistic goal to accomplish.

Thirdly, within the subsidiary or branch, management must assure that all employees understand the cultural habits, beliefs and strategies of action. Only if all groups acquire the corporate culture will it be possible to have project teams from various parts of the organization work effectively with one other. Hence, engineers, designers, production workers and accountants must know and understand each other's behaviors, nonverbal cues and habits. All the above show that training may very much depend upon the culture of which the organization is a part. Moreover, certain training methods (e.g., lecture, hands-on, on-the-job vs. off-the-job) and content (theoretical, skills practice, problem solving techniques and access of relevant knowledge) may be more successful in one cultural environment than another. Successful technology and organizational development make the coordination of training and culture into one strategy a necessity.

Theoretical Implications

It seems assured that future research on technology adoption and organizational adaptation cannot ignore the model developed here, despite its complexity. We still have far to go to develop the issues presented here before we can address the questions: How do cultural stability and training affect a firm's technological adoption and organizational adaptation efforts? Can increased tolerance for cultural diversity of a firm's workforce and its effective training (job skills and culture) help improve the firm's R&D and innovation efforts? Since these questions are central to understanding technology adoption and organizational adaptation and also technology's effective use in organizations, these issues should be a major agenda item for future research.

Studying training within a vacuum is of little help when trying to advance our knowledge about organizations and their processes. Perrow (1986, chap. 2) has argued succinctly that most organizational theories and proposed systems models ignore the complexity of the processes to be accounted for with such conceptual pieces. This paper responds by suggesting that a multidisciplinary approach is needed to study the complex organizational processes involved in deciding to adopt a technology and, subsequently, go through organizational adaptation.

The theory outlined in this paper also suggests that an exchange of ideas about technological innovation and training needs to be encouraged by providing publication outlets with a multidisciplinary perspective. Up to this point, technological adaptation and human resource management has become a fragmented discipline rooted mostly in one traditional discipline (e.g., sociology or management) instead of spanning several disciplines. One reason may be that the publication of such multidisciplinary work is quite difficult. This in turn prevents most researchers from performing work with a multidisciplinary framework, since most people do not want to put their careers at risk.

At this point, very little is known on how culture and training may together affect the outcome of technological adoption and organizational adaptation. Instead, research has concentrated on little fragments of the overall puzzle, with some notable exceptions (e.g., Hofstede, 1984). Furthermore, very rarely has research about technology concentrated on human resources and training (e.g., Gattiker, 1988c; Gattiker & Larwood, 1990). A more serious omission may, however, be the fact that up to now no attempt has been made to integrate the culture-training-technology angle as done in this paper. It is hoped that the theory outlined herein will spark some new excitement and ideas into this research domain.

Issues in the 1990s

Numerous issues arise from this paper that must be addressed by employers, employees, union representatives and public policy makers to assure the successful management of adaptation by organizations in various countries. For example, as discussed earlier, research indicates that individuals do not necessarily invite change and adaptation in their work environment caused by technology adoption. Nonetheless, such desire for stability may in part be determined by environmental factors. Resistance toward technological change may have been reinforced by undesirable occurrences negatively affecting public opinion.

For instance, political and social consequences of nuclear power plants have gained substantial media (and thus public) attention since the Three Mile Island and 1987 Chernobyl accidents occurred. Additionally, Nigeria has rejected hazardous waste from Italy and other European countries for storage in 1988. This in turn created anxiety and public awareness of possible dangerous side-effects caused by technological advancement. Hence, the public may swing away from inviting and accepting change and instead welcome stability when they see that change and innovation have led to some outcomes that are not very desirable.

Research should address the potential linkage between public opinion about innovation and technology and its impact upon resistance by employees toward organizational adaptation. Another issue requiring attention is how cultural tolerance and understanding may be improved. For instance, recently some major U.S. business schools have started to consider requiring proficiency in at least one foreign language from their graduates. The issue is the fact that knowing a foreign language may not automatically increase one's knowledge and understanding of another culture as is often implied.

Our research agenda should include further work to test if learning and acquiring a foreign language early in one's formal education and later spending a year abroad attending school/university may be a better alternative than acquiring a foreign language during graduate education without experiencing total immersion into the culture. Managers and employees with foreign language skills may not be more tolerant to cultural diversity than their colleagues. Additionally, an increase in the understanding and tolerance of multi-culturalism within one's own country may be another important step on the road to success. Unfortunately, in most countries such understanding and tolerance is hard to find.

The above suggests that managers, researchers and public policy makers must grapple with these issues to increase our understanding of the processes involved with successful technology adoption and organizational adaptation dealing with cultural stability and training of various degrees and shades. As this direction is pursued further, new theory and practical insights will be gained that may suggest not only revisions to managerial practice, but also to our general approach to work, training and technology adoption.

ACKNOWLEDGMENT

The author would like to thank Valorie Hoye and Brenda McPhail for their editorial assistance in preparing this manuscript, and my colleague Geoff England for his help with the legal arguments made in this manuscript. Financial support for this research project was provided in part by the Social Sciences and Humanities Research Council of Canada, Contract No. 492-85-1022, by a grant from the Canada Employment SEED Program, Contract No. 083080 and the Burns Endowment Fund, School of Management, The University of Lethbridge. Different portions of this paper were presented at the European Group of Organization Studies bi-annual conference in Antwerp (1987) and the High Technology Conference in Boulder (1988). The views expressed in this paper are the author's own and are not necessarily shared by these organizations.

NOTES

1. Technological innovation implies change in work structure and content. Innovation as used in this paper includes both process and product innovation leading to a reorientation of production facilities and production process improvements applying and integrating new technology into the work process. More specifically: "Technological innovation should be understood to be a technology-based process or the product of such a process that is the result of the efforts or activities of an individual, group and/or organizational system, and which represents a departure from the previous state and which may facilitate more effective resource allocation" (Gattiker, 1990a).
2. Training in an organizational context may be defined as any organizationally-initiated procedures that are intended to foster learning among organizational members. Learning, similarly, may be thought of as a process by which an individual's pattern of behavior is altered in a direction that contributes to organizational effectiveness (Hinrichs, 1976).
3. What the term theory means is unclear when looking at the social sciences literature (e.g., Hage, 1972; Homans, 1980). In this context a theory is meant to have a set of propositions and theoretical statements, each stating a relationship between its concepts. The term theory as used in this paper does, however, not only contain concepts and statements, but also outlines the theoretical and operational linkages between different elements and dimensions of the theory.
4. Cognitive ability as used in this context is synonymous with the individual's abilities for performing controlled processes. A person's efficiency in learning new skills, unfortunately, is often dependent upon the information processes that are needed to perform a given task. Automatic information processes are characterized as fast, effortless (from a standpoint of allocation of cognitive resources), and unitized (or proceduralized) in such a way that they may not be easily altered by a subject's conscious control; they may often allow for parallel operation with other information processing components within and between tasks. Automatic processes are operations that are developed only through extensive practice under consistent conditions, and include skilled behaviors as diverse as typing and skiing. As these processes become automatic, the cognitive or attentional resources devoted to the task are reduced. In contrast, controlled processes are necessary when task requirements are novel, and when the subject may not be able to internalize the consistent aspects of the task. Controlled processing is typically slow and difficult because performance is limited by the amount of cognitive resources available to the individual. An example of an activity requiring controlled processing might be making a decision to stop production in a plant due to severe quality problems, a resource intensive task which does not allow for much automatic processing (Ackerman, 1987). However, often a task may be a mixture of controlled and automated processes. For instance, bringing a new high technology product to market using the appropriate marketing techniques requires some automated processes (e.g., advertising strategy and distribution

channels to be used) while assessing the market potential and consumers' receptiveness is a controlled process (drawing the right conclusions from the test market results).

5. Cultural stability as used in this paper means the repetition and elaboration of old cultural "themes" and a single or major interpretation of the core culture. Stability can result in repetition and elaboration of old models, thereby increasing hostility to change and innovation. In contrast, limited stability results in cultural diversity and ferment seems likely to favor innovation, technological adaptation and tolerance of pluralism, thereby enhancing risk taking. Thus, national culture may not in itself be decisive in determining the degree of innovation, but instead, the attitudes held by rulers and elites toward cultural diversity may have more influence. If tolerance toward cultural diversity is great (e.g., minorities, ethnic groups, underground culture, pop art/music and religions), supporting or at least tolerating innovation and change instead of suppressing it, innovation and adaptation are more likely to occur (Goldstone, 1987).

6. Hage (1972, pp. 99-101) divides the operational linkage into four parts: First, there is the basic form such as linear, curvilinear and power (power is a coefficient that is changing); second, the direction of the operational linkage, which is either positive or negative; third, the coefficients and fourth the upper and lower limits. Thus, a correlation that is positive, ranging from 0.0 - 1.0 may have a lower and upper limit. To illustrate, technological adoption and subsequent organizational adaptation may affect grievance rates from a certain take-off point. Moreover, there may be an upper limit beyond which technological adaptation may no longer affect grievance rates.

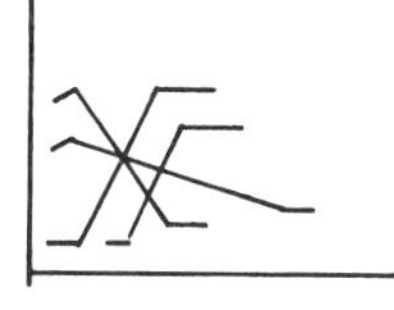

Linear correlation with different coefficient and lower/upper limits.

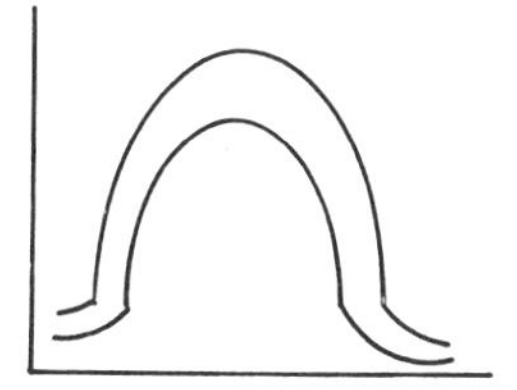

Curvilinear correlation with different coefficients and upper/lower limits.

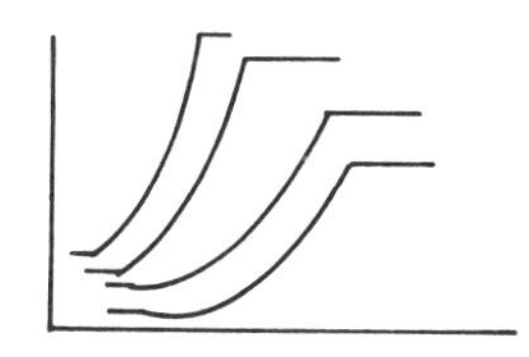

Power correlation with different coefficients and upper/lower limits.

REFERENCES

Ackerman, P.L. (1987). Individual differences in skill learning: An integration of psychometric and information processing perspectives. *Psychological Bulletin, 102,* 3-27.

Adams, L.T., Perfetto, G.A., Yearwood, A., Kasserman, J., Bransford, J.D., & Franks, J.J. (1985). *Facilitating access.* Unpublished manuscript, Vanderbilt University.

Astley, W.G., & Van de Ven, A.H. (1983). Central perspectives and debates in organization theory. *Administrative Science Quarterly, 28,* 245-273.

Attewell, P. (1987a). Numerical control machining and the issue of de-skilling. *Work and Occupations, 14,* 452-466.

________(1987b). Big brother and the sweatshop: Computer surveillance in the automated office. *Sociological Theory, 5,* 87-99.

Betriebsrat, Max-Planck-Institut fuer Bildungsforschung. Employee Works Council, Max Planck Institute For Educational Research. (1988). *Entwurf einer Betriebsvereinbarung zu individueller Datenverarbeitung im Max-Planck-Institut fuer Bildungsforschung zu Berlin.* [Outline of an employment agreement between Works Council and Employer For individual data processing on the Max Planck Institute for Educational Research]. Berlin: The Author.

Bikson, T.K., & Gutek, B.A. (1983). *Training in automated offices: An empirical study of design and methods* (Report No. WD-1904-RC). Santa Monica, CA: The Rand Corporation.

Braverman, H. (1974). *Labor and monopoly capital.* New York: Monthly Review Press.

Brumlop, E., & Juergens, U. (1986). Rationalization and industrial relations: A case study of Volkswagen. In D. Jacobi, B. Jessop , H. Kastendiek, & M. Regini (Eds.), *Technological change, rationalization and industrial relations* (pp. 73-94). New York: St. Martins Press.

Canada Labour Code, R.S.C. (1970). c.L-1 as am. Section 59.7-60.31.

Dooling, D.J., & Lachman, R. (1971). Effects of comprehension on retention of prose. *Journal of Experimental Psychology, 88,* 216-222.

Dweck, C.S. (1986). Motivational processes affecting learning. *American Psychologist, 41,* 1040-1048.

Fligstein, N. (1987). The intraorganizational power struggle: Rise of finance personnel to top leadership in large corporations, 1919-1979. *American Sociological Review, 52,* 44-58.

Francis, A. (1986). *New technology at work.* Oxford: Clarendon Press.

Gattiker, U.E. (1984). Managing computer-based office information technology: A process model for management. In H.W. Hendrick & O. Brown, Jr. (Eds.), *Human factors in organizational design* (pp. 395-403). Amsterdam: Elsevier.

________(1987a). *Identifying efficient training methods for teaching microcomputer skills to adults: Academic ability and learning performance.* Paper presented at the International Conference on Computer-Aided Learning in Post-Secondary Institutions, Calgary, Canada.

________(1987b). *Teaching microcomputer skills to adults: The effects of gender on individuals with different levels of academic ability.* Paper presented at the Giftedness of Girls and Women conference, Lethbridge, Canada.

________(1987). *Technological innovation and strategic human resource management: Developing a theory.* Paper presented at the 8th EGOS Colloquium, Antwerp, Belgium.

________(1988a). End-user computing in organizations: Facing the challenge to remain competitive. *Canadian Information Processing Society Review, 8*(6), 8-10.

________(1988b). Technological adaptation: A typology for strategic human resource management. *Behavior & Information Technology, 7,* 345-359.

________(1988c). Computer end-users: The impact of their beliefs on subjective career success. In U.E. Gattiker & L. Larwood (Eds.), *Technological innovation and human resources: Strategic and human resources issues* (Vol. 1, pp. 161-185). New York: DeGruyter.

________(1990a). Technological adaptation in organizations. Beverly Hills: Sage Publications.

________ (1990b). Individual differences and acquiring computer literacy: Are women more efficient than men? In U. E. Gattiker & L. Larwood (Eds.), *Technological innovation and human resources: End user training* (Vol. 2, pp. 141-180). New York: DeGruyter.

Gattiker, U.E., & Larwood, L. (Eds.). (1990). *Technological innovation and human resources: End user training* (Vol. 2). Berlin & New York: DeGruyter.

Gattiker, U.E., & Paulson, D. (1987). Testing for effective teaching methods: Achieving computer literacy for end-users. *INFOR Canadian Journal of Information Systems and Operations Research, 25,* 256-272.

Gick, M.L., & Holyoak, K.J. (1983). Schema induction and analogical transfer. *Cognitive Psychology, 15,* 1-38.

Goldstone, J.A. (1987). Cultural orthodoxy, risk, and innovations: The divergence of east and west in the early modern world. *Sociological Theory, 5,* 119-135.

Gregorc, A.F. (1982). *An adult's guide to style.* Maynard, MA: Gabriel Systems.

Hage, J. (1972). *Techniques and problems of theory construction in sociology.* New York: Wiley.

Hall, R.H. (1986). *Dimensions of work.* Beverly Hills, CA: Sage.

Hinrichs, J.R. (1976). Personnel training. In M.D. Dunnette (Ed.), *Handbook of industrial and organizational psychology* (pp. 829-860). Chicago: Rand-McNally.

Hofstede, G. (1984). *Culture's consequences.* Beverly Hills, CA: Sage.

Homans, G.C. (1980). Discovery and the discovered in social theory. In H.M. Blalock (Ed.), *Sociological theory and research* (pp. 17-22). New York: Free Press.

Jones, J.W., & Lavelli, M.A. (1986). Essential computing skills needed by psychology students seeking careers in business. *Journal of Business and Psychology, 1,* 163-167.

Klein, D. (1966). Some notes on the dynamics of resistance to change: The defender role. *Concepts for social change.* Washington, DC: National Training Laboratories, Cooperative Projects for Educational Development Series, 1.

Kohn, M.L., Schooler, C., Miller, J., Miller, K.A., Schoenbach, C., & Schoenberg, R. (1983). *Work and personality: An inquiry into the impact of social stratification.* Norwood, NJ: Ablex.

Kraft, P., & Dubnoff, S. (1986). Job content, fragmentation, and control in computer software work. *Industrial Relations, 26,* 184-196.

Kupfer, A. (1988, September 26). Managing for the 1990s. *Fortune, 118*(7), 44-47.

Larwood, L. (1984). *Organizational behavior and management.* Boston: Kent.

Leavitt, H. (1964). *New perspectives in organizational research.* New York: Wiley.

Lepper, M.R. (1985). Microcomputers in education. *American Psychologist, 40,* 1-18.

Miles, R., & Snow, C.C. (1978). *Organizational strategy structure and process.* New York: McGraw-Hill.

Osborn, R.N., Strickstein, A., & Olson, J. (1988). Cooperative multinational R & D ventures: Interpretation and negotiation in emerging systems. In U.E. Gattiker and L. Larwood (Eds.), *Technological innovation and human resources: Strategic and human resources issues* (Vol. 1 pp. 33-54). New York: DeGruyter.

Osterloh, M. (1986). Zum Problem der rechtzeitigen Information von Arbeitnehmervertretern in Betriebs- und Aufsichtsraeten. [Regarding the problem of timely information for employees' representatives in employee works councils on supervisory boards]. *Arbeit und Recht,*[Labor and Law], *34,* 332-340.

Osterman, P. (1984). White-collar internal labor markets. In P. Osterman (Ed.), *Internal labor markets* (pp. 163-189). Cambridge, MA: MIT Press.

________(1986). Choice of employment systems in internal labor markets. *Industrial Relations, 26,* 46-67.

Perrow, C. (1986). *Complex organizations: A critical essay* (3rd ed.). New York: Random House.

Perry, L.T., & Sandholtz, K.W. (1988). A "liberating form" for radical product innovation. In U.E. Gattiker & L. Larwood (Eds.), *Technological innovation and human resources: Strategic and human resources issues* (Vol. 1, pp. 9-31). New York: DeGruyter.

Pfeffer, J., & Cohen, Y. (1984). Determinants of internal labour markets in organizations. *Administrative Science Quarterly, 29,* 550-572.

Piore, J.J. (1986). Perspectives on labor market flexibility. *Industrial Relations, 25,* 146-166.

Piore, M.J., & Sabel, C. F. (1984). *The second industrial divide: Possibilities for prosperity.* New York: Basic Books.

Scott, W.R. (1981). *Organizations: Rational, natural, and open systems.* Englewood Cliffs, NJ: Prentice-Hall.

Shaiken, H., Herzenberg, S., & Kuhn, S. (1986). The work process under more flexible production. *Industrial Relations, 25,* 167-183.

Snow, R.E. (1986). Individual differences and the design of educational programs. *American Psychologist, 41,* 1029-1039.

Spenner, K.I. (1983). Deciphering Prometheus: Temporal change in the skill level of work. *American Sociological Review, 48,* 824-837.

Stark, D. (1986). Rethinking internal labor markets: New insights from a comparative perspective. *American Sociological Review, 51,* 492-504.

Swidler, A. (1986). Culture in action: Symbols and strategies. *American Sociological Review, 51,* 273-286.

Tanaka, H. (1988). *Personality in industry: The human side of a Japanese enterprise.* London: Francis Pinder.

UBS. (1988). UBS business facts and figures. July: 10-11.

Van Houton, D.R. (1987). The political economy and technical control of work humanization in Sweden during the 1970s and 1980s. *Work and Occupations, 14,* 483-513.

Verdin, J. (1988). The impact of information technology on human resource managers. In U.E. Gattiker & L. Larwood (Eds.), *Technological innovation and human resources: Strategic and human resources issues* (Vol. 1, pp. 143-159). New York: DeGruyter.

Wilpert, B. (1986). Leadership and decision making in introducing new technologies. In J. Misumi (Chair), *Cross-cultural and interdisciplinary perspectives of leadership and organizational development.* Symposium conducted at the 21st International Congress of Applied Psychology, Jerusalem.

THE AUTHORS

Marietta L. Baba is Associate Provost and Director of International Programs, and Professor of Anthropology at Wayne State University. An industrial anthropologist, Dr. Baba's main research interests are in the areas of technological innovation and organizational evolution. Dr. Baba received her Ph.D. in Physical Anthropology from Wayne State University in 1975. Her backround includes a decade of research in the field of molecular evolution and studies in primate ethology at the Yerkes Regional Primate Center. Recently she served as Co-Principal Investigator (with Herman Koenig of Michigan State University) on a three-year NSF-sponsored study of university-industry relations in Michigan. Based on this research, Dr. Baba has published papers on the historical development of university-industry linkage types, university-industry linkage in the field of anthropology, and university innovation and environmental change. Dr. Baba's current research focuses on the structure and evaluation of international joint venture organizations, which has led to a new, patented ethno-historical mapping method for analyzing partnership organizations over time. In addition to teaching, thesis direction, and administrative duties at Wayne State, Dr. Baba is a consultant at General Motors Research Laboratories. She is immediate Past President of the National Association for the Practice of Anthropology (NAPA), and founded NAPA's Business and Industry Group.

Stephen R. Barley is Assistant Professor of Organizational Behavior at the New York State School of Industrial and Labor Relations, Cornell University. His current research interests include occupational and organizational implications of technical change, organizational culture and the industrialization of science. He received his Ph.D. in organization studies from the Massachusetts Institute of Technology. He has published papers in *Administrative Science Quarterly, Research in Organizational Behavior, Research in the Sociology of Organizations,* and *Contemporary Ethnography.*

Janice M. Beyer received both the M.S. and Ph.D. degrees in organizational behavior from the New York State School of Industrial Labor Relations at Cornell University. She has served on the faculty of the School of Management at the State University of New York at Buffalo as Assistant Professor (1973-1978), Associate Professor (1978-1981), and Full Professor (1981-1986). From 1986 to 1988 she was Professor of Management at the Graduate School of Business at New York University. She is currently the Rebecca L. Gale Regents Professor in Business at the University of Texas at Austin.

Dr. Beyer's research interests center on issues of organizational culture and change. Her publications include a co-authored research monograph entitled *Implementing Change,* and over 50 articles on such topics as organizational design, interorganizational relations, rites and ceremonies in organizations, and the utilization of organizational research. Dr. Beyer is currently President-elect of the Academy of Management. From 1985-1988 she was Editor of the *Academy of Management Journal.* She has held numerous other offices in the Academy of Management and other professional associations, and has served as a frequent reviewer for other professional journals and for funding agencies of the U.S. and Canadian governments.

Robert D. Bretz, Jr. is Assistant Professor of Personnel and Human Resource Studies in the New York State School of Industrial and Labor Relations at Cornell University. He received his M.B.A. and Ph.D. in Personnel/Human Resource Management from the University of Kansas. Before joining the faculty at Cornell, Professor Bretz served as a research assistant at the Institute for Public Policy and Business Research at the University of Kansas. While at the University of Kansas, he also earned a School of Business Dissertation Fellowship and the Graduate Business Council Outstanding Educator Award. His primary research interests include the individual job search and choice process, staffing and selection issues and the potential causes and effects of work force homogeneity.

M. Ronald Buckley is Assistant Professor of Management at the University of Oklahoma. He received a Ph.D. in industrial/organizational psychology from Auburn University. Dr. Buckley's research interests include employment interviewing, performance appraisal and feedback, and construct validation. His research has appeared in such journals as *Journal of Applied Psychology, Organizational Behavior and Human Decision Processes, Academy of Management Review, Journal of Management* and *Applied Psychological Measurement.*

Stephen J. Carroll received his Ph.D. from the University of Minnesota. He has been a faculty member at the college of Business and Management in the University of Maryland since 1964. He is presently affiliated with the Dingman Center of Entrepreneurship and the Center for Innovation in the College. He is the coauthor of nine books, four monographs, and many published articles and research papers in leading academic journals. Prof. Carroll has served as a consultant to more than thirty leading U.S. companies and government organizations.

Wayne F. Cascio earned his B.A. degree from Holy Cross College in 1968, his M.A. degree from Emory University in 1969, and his Ph.D. in industrial/organizational psychology from the University of Rochester in 1973. Since that time he has taught at Florida International University, the University of California-Berkeley and the University of Colorado-Denver, where he is presently Professor of Management.

Dr. Cascio is a Fellow of the American Psychological Association, a Diplomate in industrial/organizational psychology of the American Board of Professional Psychology, and a member of the Editorial Boards of *Human Performance, Organizational Dynamics* and the *Academy of Management Review*. He has consulted with a wide variety of organizations in both the public and private sectors on personnel matters, and he periodically testifies as an expert witness in employment discrimination cases. Dr. Cascio is an active researcher, and is the author of four books on human resource management.

George F. Dreher is Associate Professor of Management at Indiana University. He received his Ph.D. in Industrial-Organizational Psychology from the University of Houston. Professor Dreher is actively involved in staffing and compensation research and has published several articles on these topics in *Personnel Psychology, Journal of Applied Psychology, Journal of Vocational Behavior, Industrial Relations, Academy of Management Journal, Academy of Management Review,* and others. He has also published *Perspectives on Employee Staffing and Selection* (with Paul Sackett). Professor Dreher is on the Editorial Board of *Personnel Psychology* and has served as ad hoc reviewer for the *Journal of Applied Psychology, Journal of Management Studies, Psychology Bulletin* and *Academy of Management Review.* Additionally, he has been actively involved in management development programs for companies including Cities Service, FMC, Natural Gas Pipeline Company of America and United States Telephone Association.

W. Gibb Dyer, Jr. is Associate Professor of Organizational Behavior at Brigham Young University. His research interests include organizational culture, innovation in organizations and the management of family owned businesses. He is on the editorial board of the *Family Business Review*.

Gerald R. Ferris is Associate Professor of Management at Texas A&M University. He received a Ph.D. in Business Administration from the University of Illinois at Urbana-Champaign. Dr. Ferris has research interests in the areas of social influence and organizational politics, performance evaluation and the employment interview. He is the author of over 40 articles, which have appeared in such journals as *Human Relations, Journal of Applied Psychology, Organizational Behavior and Human Performance, Personnel Psychology, Academy of Management Journal* and *Academy of Management Review*. Dr. Ferris is coeditor of *Method & Analysis in Organizational Research, Personnel Management, Current Issues in Personnel Management* and *Human Resource Management: Perspectives and Issues*. He also serves as coeditor of the annual series *Research in Personnel and Human Resources Management*.

Gerald Fryxell is Assistant Professor of Management at the University of Tennessee-Knoxville. Prior to earning his Ph.D. in Strategic Management at Indiana University in 1986, he was a Medical Technologist (ASCP), and an instructor at the University of Wisconsin-Superior. His current research interests are in the areas of corporate culture, innovation, strategic management and organizational justice.

Urs E. Gattiker is Assistant Professor of Organizational Behavior and Technology Management at the School of Management, University of Lethbridge in Alberta, Canada. He received his Ph.D. in organization and management from the Claremont Graduate School in 1985. Dr. Gattiker is co-editor of the De Gruyter biannual Series on *Technological Innovation and Human Resources* and is Director of the *Technology Assessment Research Unit*, University of Lethbridge. His area of specialization and prior publications include career development, end-user computing, technological and organizational assessment, and management consulting. He is currently Chair of the Academy of Management's Research Methods Division.

Luis R. Gomez-Mejia, Ph.D. in Industrial Relations, University of Minnesota, is a full Professor of Management at Arizona State University. Previously, he was a codirector of the High Technology Management Research Center at the University of Colorado. He has had over four years of full-time field experience at Control Data Corporation and has also been a consultant on human resource problems to numerous high technology fims. Dr. Gomez-Mejia has over 50 publications appearing in *Administra-*

tive Science Quarterly, Academy of Management Journal, Strategic Management Journal, Industrial Relations, Personnel Psychology, and others.

Ralph C. Hybels is a Ph.D. candidate in organizational behavior at the New York State School of Industrial and Labor Relations, Cornell University. Prior to his enrollment at Cornell, he worked as a technical recruiter in agencies in the Boston area. His dissertation research focuses on the determination of organizational strategies through interoccupational politics and top executive turnover in high technology.

Mariann Jelinek is Associate Professor of Management Policy at the Weatherhead School of Management, Case Western Reserve University, where she was tenured in 1987. Previously, she held appointments at the State University of New York-Albany, McGill University, the Amos Tuck School at Dartmouth College, Worcester Polytechnic Institute and Bentley College. She was educated at the University of California-Berkeley, where she received her Ph.D. degree; and at Harvard Business School, where she earned her D.B.A.

Dr. Jelinek's teaching assignments include Policy and Advanced Manufacturing Technology and Corporate Strategy. Research and consultation interests center on effective management of innovation, strategic change and manufacturing technology. Dr. Jelinek has published six books, including *Institutionalizing Innovation,* and more than 20 articles, among them "Plan for Economies of Scope," an article with Joel Goldhar, which appeared in the *Harvard Business Review.* A book on the management of high technology firms, *The Innovation Marathon* (with C.B. Schoonhoven), was published in 1989.

Harvey Kolodny is currently Professor of Organization Behavior at the Faculty of Management, University of Toronto. His special fields of interest are organization design, organization theory, the quality of working life and the management of technology. He has worked in, researched, taught about and consulted a variety of Canadian and American organizations on the specific topics of design and management of complex organizations, particularly matrix designs, sociotechnical system design of new and existing factories, and general job and organization design methods that lead to improvements in the quality of people's working lives. Dr. Kolodny received his doctorate in organization behavior from the Harvard Business School. He has an M.B.A. degree from the University of Sherbrooke (Quebec) and an engineering degree from McGill University. Dr. Kolodny's current research includes a comparative international project (Sweden, France and Canada) to examine the skills required by managers and engineers to effectively implement new technology into their organization's structure and processes.

Michael Lawless, Ph.D. in Strategic Analysis, Anderson Graduate School of Management, UCLA is Associate Professor at the Graduate School of Business, University of Colorado at Boulder, and Director of the High Technology Management Research Center. His experience includes staff and consulting positions at the RAND Corporation and System Development Corporation. Professor Lawless' research on competitive strategy, innovation, and strategic management of technology has been published in *Strategic Management Journal, Management Science, Human Relations, Journal of Management Omega,* and others. Dr. Lawless is Associate Editor of the *Journal of High Technolody Management Research* and an editorial board member at the *Journal of Management.*

Robert A. Page, Jr. is currently a Ph.D. student at the University of California, Irvine, studying organizational behavior. He received a Master of Organizational Behavior degree from Brigham Young University in 1986, and a B.S. degree in Industrial and Labor Relations from Cornell University in 1983. Beyond innovation, his research interests include organizational theory, organizational culture and motivation.

Joseph G. Rosse is Assistant Professor of Strategy and Organization Management at the University of Colorado at Boulder. He received his Ph.D. in industrial/organizational psychology from the University of Illinois in 1983, and now teaches, conducts research, and consults in the areas of employee staffing and retention and employee counter-productive behaviors.

Claudia Bird Schoonhoven is Professor of Organization and Management, San Jose State University, School of Business. Her Ph.D. and M.A. are from Stanford University, in organization theory and behavior. A specialist in the management of innovation, organizational design and performance, currently she is the recipient of research grants to investigate differential mortality rates among new technology ventures, and the impact of incubator region characteristics on survival of new semiconductor industry ventures. Recent publications include *The Innovation Marathon,* and "Dynamic Tension in Innovative High Technology Firms" in *Managing Complexity in High Technology Industries* (M.A. Von Glinow and S.A. Mohrman, Editors), both coauthored with M. Jelinek.

Donald L. Sexton is the William H. Davis Professor in the American Free Enterprise System in the College of Business, the Ohio State University.

Prior to joining Ohio State, he was the Director of the Center for Entrepreneurship and the Caruth Professor of Entrepreneurship in the Hankamer School of Business at Baylor University. Dr. Sexton holds a BS in Math/Physics from Wilmington College and an MBA and Ph.D. from the Ohio State University. Prior to joining academe, he spent 18 years in industry. He has coauthored a number of books, and served on the editorial review boards of the *Journal of Small Business Management,* the *Journal of Business Venturing* and the *Academy of Management Review*. He is also the Entrepreneurship Series Editor for Macmillian.

Mark S. Turbin earned a Master's degree in industrial/organizational psychology at the University of Illinois, where he also did further work toward a Ph.D. while teaching and consulting for the University and several other organizations. He then joined the management faculty at the University of Wyoming. His teaching and research have included the areas of organizational behavior, motivation, measurement, performance appraisal, personnel selection and retention, organizational research and consulting and is a research associate at the Institute for Research on Social Problems in Boulder, Colorado.

D.D. (Don) Warrick is a Professor of Management at the University of Colorado at Colorado Springs, where he is the Director of the Creative Management Center. He is also the President of the Warrick Agency Training and Development Company and has been a consultant or trainer for over 400 organizations. Dr. Warrick has received many national awards for his achievements, including being named the Outstanding Organization Development Practitioner of the Year (1982) and the Outstanding Human Resources Professional of the year (1984) by the American Society for Training and Development. He has served as the Director of the Organization Development Division of the Academy of Management and the editor of the *Academy of Management Organization Development Newsletter,* and is on the editorial board of *Group and Organization Studies.*He is the author or co-author of four books.

Jan Zahrly is associate professor of management and organization sciences at Old Dominion University, where she teaches strategy and organization theory. Her research interests include the problems of new organizations, agency theory and the effects of long term versus short term strategies for corporations. Dr. Zahrly has presented her research at national and regional conferences and has published articles in the *Journal of Organizational Behavior, Psychological Reports,* and others. She received her Ph.D. from the University of Florida (1984).

INDEX